Beginner's Money, Saving and Investing

Discover New Ideas And Let's Get Started Saving And Growing Your Money

EMIL COLLINS

COPYRIGHT

TABLE OF CONTENT

Just For You!

A FREE GIFT TO OUR READERS

Ready-to-use templates that you can download and start mind mapping with right away! **Visit this Link**

https://infobusiness001us.activehosted.com/f/1

PREFACE

Are you looking for a way to overcome your financial worries by discovering fool-proof strategies for saving money and starting your own Business? This is the book for you!

The financial crisis and money related woes aren't something that any of us ever want to experience. Yet, it is almost impossible with the unexpected nature of life to properly prepare for a future where everything is nothing but perfection.

As long as you have Money, as long as you earn Money and your family members rely on your for financial support, you are bound to face a financial crisis at one point or another! This is pretty much an unavoidable truth of life. What you can do, though, is immediately start to smart save your Money, invest them in a business to ensure a safe and secure future and prepare yourself for unforeseen emergencies to the best of your abilities.

This book titled **"Beginner's Money, Saving And Investing: Discover New Ideas and Let's Get Started Saving and Growing Your Money"** has written from the ground up to help you do that and more. The core ideas that you will discover in this book include;

- Understand the fundamental concepts of Money and why it is important to start saving up for the future.
- Understand the importance of setting up the Financial Goal.
- Know what the Long Term, Mid Term and Short Term Financial Goals are and how you can properly implement them in your life.
- Explore various strategies through which you can start saving Money right away.
- Know about some of the modern and most up-to-date methods of Earning Money, both physically and online.
- Get to know the core steps that are involved in establishing a successful business.
- Fully understand how to research your market before starting a business, with a basic step by step breakdown of the process.
- Discover how to create the definitive business plan for success.
- Get to know how you can effectively grow your business and increase the rate of your organic customers.
- Discover the importance of choosing the location of your Business and learn more about the decision-making process to choose the perfect place and considerably much more!

If you have any last-minute hesitations in purchasing this book, leave them right now, as this is essentially a goldmine of information for those who are looking to take their first step towards the financially independent life.

So, if you want to discover how to save money right now, start out your own business and start paving your path towards the life that you seek, buy this book right now. Take the first step towards changing the whole outlook of your financial structure!

INTRODUCTION

If recent statistics are to be believed, then according to a finding established by Employee Benefit Research Institute, just a shocking 14% of the total population of America claim that they are "Confident" to have enough money in their savings for a relaxing and worry-free retirement. A whopping 30% claims that they have less than $1000 in their savings account, and even worse, the remaining 56% confess that they never even calculated how many savings they would need for retirement!

To make matters worse, Creditloan.com released a statement claiming that a typical American usually ends up having their first debt as early as being in High-School!

Feeling worried Yet? Well, don't be! Let the current grim scenario be a lesson for you and inspire yourself to take the first step towards starting a savings account.

If you are a person who earns money, you have other members of your family and dependents relying on you for economic support, then experiencing an ongoing financial crisis at any stage of life is almost given! No matter how much you try, you will never be able to fully prepare yourself for a medical emergency, a sudden and unexpected trip to your parents, a major expenditure in building repairs and a rise in rental fees...the just goes on and on.

I mean, who knew that the Pandemic such as the COVID-19 would hit and bring the whole world to its knees, right? If you didn't save money before this crisis, there might be the chances of your regret and you might be thinking that why didn't you start saving in the first place!

In short, if you have money, then you are bound to face money-related issues in your life.

There was a time when even I was had to endure severe money management problems in my life. I had to live on a very strict and suffocating budget and it prevented me from doing anything that I wanted to. I was essentially living paycheck to paycheck. Things got even harder for me when I started a family. All the major expenses seemed to exponentially double down. It was becoming excessively hard for us to cope with the economic inflation.

So, believe me, when I say this if you are reading this book, chances are that you are going through the similar financial crisis. I can totally relate and understand just how much frustrated you are.

The good news is that this is a very much solvable problem if you are smart enough and know how to properly manage your revenue streams. But there is another thing that is very important if you want financial independence. That is to have your very own business.

I learned this the hard way, but you don't have to. Once I hit total financial rock bottom, I decided to strictly take control of my financial efforts. My family wholeheartedly supported me. I started researching various methods of saving money and about how an individual can use the saved money to invest in a smart business plan. It took me a while to get things right. Still, with our growing passion for saving money together and my internal desire to start a business, I was able to save enough money to fully support my family and our life and launch a successful Kindle Publication that has been supporting us ever since!

It took me a while, and it wasn't easy, but the dedication and effort all paid off in the end. And it all started with saving money.

Saving money is one of the most important aspects of accumulating capital and ensuring a stable financial future. Saving money is a way out of life's uncertainties and allows you to live a more fulfilling life. Putting money aside in a structured way will help you avoid several pitfalls and challenges in life. It will help you in your hour of need and ensure that your family has something to fall back on in the disaster.

Consequently, once you have enough capital, you must get out of your comfort zone and start exploring some low-cost business ventures. Running your own business, means facing many challenges. However, for many people, the benefits of running a business greatly outweigh the challenges and difficulties. According to a recent survey conducted by Guidant Financials Small Business Trends Alliance, 76 percent of those who decide to run their own small business are "satisfied" or "very happy" with their decision.

Contrary to popular belief, you don't need to have an entrepreneurial spirit to run a business! Running a business can be exciting for many people for various reasons. There are many advantages to starting your own business whether you want a more flexible lifestyle, or are willing to experiment, or simply want more leverage over your job and earning potential.

If you have steady capital and savings, then there's no reason as to why you should not try your hands 0ut on a business. Especially, in a world where anyone can start a digital business with almost minimal to no investment!

Whether you are a family man, who is financially struggling or are an individual who wants to know more about the importance of saving money so that you can prepare yourself for the future. This is the perfect book for you! The book has been very meticulously written and crafted so that anyone can pick it up, even if you are a complete beginner!

Following the theories, strategies, and ideas that are shared within this manuscript, you will:

- Be able to attain full financial flexibility and independence.
- Attain peace of mind, knowing that you are financially confident enough to support your family and save enough for a retirement.
- Know exactly how you can save money to fulfill your dreams while not going bankrupt.

- Be able to enjoy all the financial rewards that come with saving money and smart investment.
- Be able to directly help others in need through your business, which will, in turn, give you a great deal of satisfaction.
- Be able to create a platform for innovation.

Beside these above, most importantly.

Be able to secure a future for you and your family. Where you can peacefully live out the rest of your life knowing that you are prepared for an emergency situations and major expenses that may come.

To sum up, this book is designed to be your go-to to solve all your finance-related problems.

For those who want to know what topics that are covered in this book, let me give you an overall overview of what you can expect:

- Understand the importance of money, know-how the concept of "Money" came to be, and why should you consider starting to save your money right now!
- Understand the importance of having a clear Savings Goal. Know the difference between long-term, midterm, short savings goals and discover clear-cut steps on how to set up and follow up on those goals.

- Explore various full-proof strategies that you can start implementing in your life right now and start saving money from almost all corners of your life.
- Discover some of the latest and greatest strategies through which people earn money both online and in real life.
- Discover the process of building up your very own business from scratch with an easy and intuitive step-by-step guide that will let you walk from developing your idea, doing the market research, creating the perfect business plan, establishing your business, and finally growing and boosting your business to the next level and above that.

That being said, let's jump into the first chapter now, where I will help you understand the fundamental concepts of money, the importance of saving money and keep exploring more as we move forward.

CHAPTER ONE

The Importance and the Value of Saving Money

Before you understand and learn to appreciate the value of money, you must first understand what money exactly is. To most of us, it is just a means of exchange, right? Well, money is much more than that.

What exactly is Money?

Money is what makes the world go round. The exchange of money for goods and services is the foundation of economies.

People used to barter to get the goods and services, they wanted before the invention of a medium of trade, aka. Money. When there was no money, two people who would have something that both parties are interested in would look at the products, come to an agreement and make an exchange.

However, early forms of bartering lack the transferability and divisibility that makes trading effective. For example, if someone has cows but needs bananas, they must find someone who has both bananas and a desire for meat. What if that person comes across someone in need of meat but lacks bananas and can only give potatoes? To obtain meat, the person must first locate someone who has bananas and desires potatoes, and so on.

Situations like these often made the process of bartering a rather complicated one. Bartering for products is exhausting, confusing, and wasteful due to the lack of transferability. But that is not the end of the problems; even if the individual meets someone to exchange meat for bananas, many bananas might not be worth a whole cow. Such a trade necessitates reaching an agreement and devising a method to decide how many bananas are worth which sections of the cow.

These issues were resolved through the use of commodity money. Commodity money is a type of a good that serves as a form of currency. American colonists, for example, used beaver pelts and dried corn in transactions in the 17th and early 18th centuries. These goods, which had widely agreed values, were used to buy and sell other items. Commodities used in trade possessed certain characteristics: they were highly desired and expensive, but they were also robust, compact, and easily stored.

A precious metal, such as gold, is another advanced example of commodity currency. Until the 1970s, gold was used to back paper money for decades. In the case of the US dollar, for example, this meant that the foreign governments might take their dollars and exchange them for gold at a predetermined rate with the US Federal Reserve. What's fascinating is that, unlike beaver pelts and dried corn, which can be used for clothes and food, gold is valuable solely because people want it. It isn't inherently useful—you can't eat gold, and it won't keep you warm at night—but the majority of people think it's amazing, and they know others do, too. As a result, gold has monetary value. As a result, gold acts as a tangible symbol of wealth based on its expectations.

This relationship between money and gold reveals, how money acquires its value—as a reflection of something precious.

The second form of money is fiat money, which does not require physical backing. Instead, the value of fiat currencies is determined by supply and demand, as well as people's belief in their worth. Since gold was a scarce resource and rapidly expanding economies couldn't always mine enough to back their currency supply requirements, fiat money formed. The need for gold to offer money value in a thriving economy is extremely inefficient, particularly when its value generated by people's perceptions.

Fiat money becomes a symbol of people's perceptions of value, serving as the foundation for why money is made. An increasing economy appears to be effective in generating other goods, beneficial to itself and other economies.

The stronger the economy, the more valuable its money would be considered (and desired), and vice versa. However, people's expectations must be matched by an economy capable of producing the goods and services that they need.

For example, in 1971, the United States dollar was decoupled from the gold standard, which meant that it was no longer redeemable in gold. The price of gold was no longer set to any dollar amount.

This meant that it was now possible to print more paper money than there was gold to back it up; the strength of the US economy backed up the dollar's value. If the economy slows, the US dollar value will fall both domestically and internationally due to inflation and currency exchange rates. The collapse of the United States' economy will plunge the world into a financial abyss, so many other governments and institutions are working hard to ensure that this does not happen.

Today, the buying power of money (not only the dollar but most currencies) determines its value solely by inflation. That is why merely printing new money would not result in a country's wealth increasing. Money is produced due to continuous interaction between true, tangible objects, our desire for them, and our abstract confidence in what has meaning. Money is important because we value it, but we desire it only because it allows us to obtain a desired good or service.

How Money Is Created?

We've spoken about why and how money, representing perceived value, is generated in the economy. Still, another significant aspect of money and the economy is how a country's central bank (the Federal Reserve or the Fed in the United States) can control and manipulate the money supply.

The Fed will, of course, print money if it wishes to increase the amount of money in circulation, maybe to stimulate economic activity. However, physical bills account for just a small portion of the money supply.

The central bank can also increase the money supply by purchasing the government fixed-income securities in the market. As the central bank purchases these government securities, it injects capital into the economy and, in essence, into the general public's hands. How does a central bank like the Fed fund this? As odd as it might sound, the central bank actually generates money and distributes it to those who sell securities. Alternatively, the Fed will lower interest rates. Which will enable banks to lend low-cost loans, credit and encouraging companies and individuals to borrow and spend.

The central bank does the reverse and sells government securities to reduce the money supply, possibly to reduce inflation. The currency used to pay the central bank is effectively removed from circulation. Keep in mind that in this case, we are generalizing to keep things easy.

A central bank cannot continue to print money indefinitely. If an excessive amount of money is released, the value of that currency will fall by the rule of supply and demand.

Remember that as long as people believe in the currency, a central bank will print more of it. However, if the Fed prints too much currency, the value would fall, as with anything with a higher supply than demand. As a result, the central bank cannot literally print money whenever it wishes.

Why Saving Money Is Important?

If you don't make much money and can barely pay your bills, the idea of saving money can seem absurd. Why bother trying to save when you just have $5 left at the end of the month?

One simple answer, because everyone has to start somewhere, right? If you keep waiting for that big break or the moment of truth to come when all your money issues go away, then you won't be able to save even a penny.

So, your best bet is to start saving right now! Believe it or not, if you start saving up now, your financial situation will change down the line. It might seem a bit difficult at first, but you will thank yourself down the road.

It will give you mental peace and make life much easier for you later on.

So why is it actually important to save money?

1. **Mental Peace**

Who hasn't woken up at 3 a.m. wondering how they were going to pay for anything they needed? If money is tight, you will be concerned about how you can pay the rent next week. If you're a little higher up the financial ladder, you may be concerned about how long you'd be able to pay your bills if you lost your work. Later in life, the money worries that keep you up at night will revolve around paying for your children's college education or getting enough money to retire.

As you save money, your financial concerns should lessen as long as you live within your means. You'll sleep easier at night, if you've already paid your rent for the next month by the first week of the next month. if you realize you can go three to six months without working, and if you have savings plans for your children's education and your own retirement that you're constantly supporting. The relief from stress that comes from saved money allows you to devote your attention to more pleasant thoughts and activities. Finding the right savings account is critical to ensuring that the money you save receives the highest interest.

2. More Control Over Life

The more money you have invested, the more leverage you have over your own fate. If your job is causing you to have a nervous breakdown, you should resign even if you don't have a new job lined up yet, and take time off to regain your sanity before looking for new work. Suppose you're tired of dangerous neighborhood. In that case, you should relocate, and you'll have enough money for a deposit on a better apartment or a down payment on a nicer home.

Become ill and need costly healthcare that your insurance does not cover. You will be able to pay for it even if you are unable to function while receiving treatment. And knowing you have choices because of the money you've saved will provide you with even more peace of mind.

No, money does not solve all problems. If you are laid off, it could take up to two years to find another job. Certain conditions will not go away regardless of how many procedures you can afford, and random crime will occur even in a supposedly safe gated community. However, having more money in the bank to deal with problems like these increases the chances of winning.

3. The Confidence To Invest

Most of us work hundreds of hours a year to earn the majority of our income. However, when you save and invest your money in the right ways, your money begins to work for you. You'll need to work less and less as your money grows, and finally, you may be able to quit working entirely.

What does it mean for your money to work for you? When you first start saving, you'll want to keep your money somewhere safe, where you can get to it quickly in case of an emergency. That means an online savings account, where you can receive 1% interest per year and not even keep up with inflation, which typically runs between 1% and 2% per year. You'll still have to pay taxes on your meager 1% profits. However, anything is preferable to receiving 0% or not saving and incurring credit card debt, which can cost you 10% to 30% in interest each year.

After you've saved three to six months' worth of expenses in your emergency fund, you can begin contributing to a tax-advantaged retirement account. That is when the magic begins to happen. You can invest in the stock market through these plans, such as a Roth IRA or 401(k). You won't have to pay taxes on your investment profits, which will allow your money to grow even faster. You contribute after-tax dollars to a Roth IRA, and everything in the account after that is yours to hold. You will contribute before-tax dollars to a 401(k), allowing you more money to spend upfront; you will pay taxes when you withdraw the money after retirement. (If you're unsure about paying taxes now or later, you can hedge your bets by contributing to both your employer-sponsored retirement account and a Roth IRA.) A standard IRA, the third option, allows you to invest before-tax dollars, much like a 401(k) (k).

If you have a high income and low expenses, you may retire in ten years. For the majority of people, it takes closer to 40 years. However, suppose you save and invest regularly, you should eventually live off the income provided by your investments. This saved money is now working for you.

The sooner you begin, the longtime growth of the small investment happens because of the magic of compounding.

4. Financial Independence

Everyone desires financial independence, but not everyone can achieve it. The major cause is that people don't understand that financial independence isn't magic that can be taken by turning the magic wand. Still, it's a process that needs commitment, dedication, and effort for a long period.

So what is financial independence? In simple words, it's how you can survive and fulfill all your needs, wants, and demands through procuring enough money. Now, remember, financial independence always works in the future. That means it requires saving money today so that you can have a better tomorrow. By better tomorrow, I mean rich and wealthy living.

Just imagine spending time with your loved ones after retiring at an early age, going on vacations, and spending time together to make unforgettable memories. Sounds like a dream, right? But it can be true if you know how to save money to be financially independent. More to this, it offers you the opportunity to choose when to work and have a day off. It allows you to follow your passion without worrying much about risks. It gives you the power to magnify financial security, as you don't depend on living paycheck to paycheck.

I've also noticed one more thing that people who ignore to think about financial freedom usually end up losing everything. Why? Because they don't save money on their favorable days. And they ignore the fact that there would be days when they won't be able to work, and the only worldly thing that would help them is their saved wealth. I'm sure nobody wants to end up living on other's support in the last few days, so it's important to start saving because it'll not only help you to live an easy life during middle age, but you'll appreciate it to the fullest even in your senior years.

5. The Freedom To Invest

"Oh! I wish I had invested a few years ago. We would have been millionaires by now. BUT son, your mom and I had no savings back then."

Well, that 'BUT is a painful excuse, I've heard from a lot of people. Remember, opportunities are given to everyone, but not everyone can achieve them. Investing in the future requires monetary stability, which is why it's necessary to start saving money from today.

When you have money, you're free to invest in things, which could benefit the long term. I have many examples to cover my statement, and I'm sure you've likely read several success stories too.

Although you could learn more about investing in my next book, that's a whole different topic. However, one way or another, it's related to saving money. So what happens when you save money? You can invest in different products, services, assets, and many other options, which could return some heavy profits over time.

Money helps you buying valuable things, which grow further in their value. In the more technical language, you can buy things that will pay you later.

6. Get Out Of Debts

Here's another obstacle that won't let you save money, and it's a real pain in the head and heart. Most of us, in search of lavish living, begin to believe in borrowing money because it has to be returned with a hefty amount.

Yes! I'm talking about loans and interest upon them. Credit cards, mortgages, and hundreds of other options are available that might satisfy you for a while or even for a couple of years. Still, they're totally back-breakers in the long run. How?

The interest rates are way too heavy to pay on the fixed borrowed amount. As a result, you're unable to save money because you have to pay the dues. The funny part is, many lending institutions even allow you to borrow more to pay the old loans. So it's like a never-ending pit where you keep on falling.

So, if you want to get out of debts, you need money for that, and I'm not talking about taking more loans, but your savings. I know it seems pretty difficult to save money when you're already paying the entire amount, but that's what I want you to be good at.

A little compromise on your lust and desires, for now, can help you save some bucks. There will soon be a day, when you have enough amount to pay off the interest and loans to live with freedom.

7. The Confidence To Buy Assets

You should know what an asset is? In the financial and accounting sphere, we use this word to mention any tangible property owned by an individual or an organization.

But apart from that bookish definition, assets could be anything that has a value, and the value could be increased or decreased.

A person could be an asset, and so does a thing.

Yes! I've talked about it in the previous subheading related to investment. I've even given you the example of buying a home and how it could be profitable (Lexico, n.d.). But that's not the only asset left in the world. There are some extremely awesome money-making assets you can invest in when you have money. Here's the list: Peer-to-peer lending

- Property rentals.
- Launching your product.
- Investing in marketing your product or services.
- Getting a savings account.
- Certificates of Deposits.
- Buying farms.
- Starting an eCommerce Business.

Additionally, there could be hundreds of more names. The more you'll add the sugar, the more it'll taste sweet: That's why saving money is essential.

Otherwise, you cannot buy assets. Once you have saved a little amount, buy assets. Even the smallest asset could grow bigger in the future.

But make sure, your assets are likely to increase. So, you have to do a little research to identify whether buying a specific asset is profitable or not. But all that can only happen if you have money in your account.

8. Tackling Emergencies

Bad times are unpredictable, and they come uninvited. Ask yourself if you're ready to handle any unwanted situations, which could only be managed with the help of money? Of course, I'm not referring to life or death, as these are natural incidents, which are surely going to take place someday. But there are times when money can help you escape many ugly situations.

Let's understand this with the most common example. You're in a hospital, and if you're given an expensive treatment, then the chances are there your illness could be cured. But what if you cannot afford it? I know there are legal ways to gather money for treatment. Still, those are secondary and likely inconvenient options. But when your bag is full, you can spend without a second thought? Why? Because your only goal would be to become healthy and get out of that hospital bed.

And here's another example. Imagine a vehicle hit your car while it was in the parking lot. Your car is seriously damaged, and you couldn't find the person responsible. What would you do? Claiming your insurance will be one way.

The second way would be taking care of the fixes by paying from your own wallet. But whatever you choose, you need to have money.

The car insurance company won't do it for free unless you are paying regular premiums. In the same way, if you take the burden of expenses on your shoulders, you still need money. Therefore, saving money is important, as it can surely help you in some of the most distressing times.

9. No Financial Trouble

Let's shed light on the recent pandemic. COVID-19 has been one of the worst pandemics since the 1920's Spanish flu. The consequences have been ugly on not only US citizens, but also the majority of the people worldwide have lost their jobs to it. But those who had savings survived in a better way as compared to those who didn't.

They were able to pay their bills, pay their rent, and arrange food from their piggy banks until the world found ways to cope with the situation.

But one thing that we all have learned and still learning that saving money can be the way to exit financial trouble in unfavorable circumstance. Those who saved money faced little or no trouble.

10. Financing Family and Social Responsibilities

When you have savings, you can always come forward to help your family, friends, and community. Just think how beautiful it would be to financially support your friend in a business, or your family member in case of any emergency, like taking care of the medical expenses and everything else.

Conversely, when you have savings, your mind is open to long-term planning. You never have to be embarrassed when your child asks you to pay the college fee which he or she ever dreamt of getting enrolled into. Plus, you're always there to support your spouse in all situations.

11. Learning to Manage Money for Business

Now, this is something unique about saving money. I enjoy this factor the most. I've observed that there are two kinds of people. One who saves and spends wildly on the things which they even don't need. And other ones, who spend it wisely, and those who are wise with their money, understand how to manage it.

And this difference between 'spending wildly' and 'spending wisely' is what makes you or breaks you. In my case, I became wise when I started saving money. I began to judge my observations. I became more thoughtful with my investments and buying assets, and I became what I am today. Although not a famous billionaire yet, still I've changed peoples' lives. And that's what I'm proud of.

So, if you wish to learn how money management can help you get your goals, start saving from now, even if there's a dollar in your pocket.

12. Peaceful Retirement

This is an important benefit. Why? Because money is needed more when you grow old and retire from your job or business. So, when you have savings, you can surely buy assets or invest in continuing the cycle.

What I suggest is to start saving money from an early age because when you retire, you'll have plenty of options to consider. And for those readers who didn't get any guidance during the youth, don't worry because you know what to do now.

Never forget that opportunities are always there. You just need to knock on the right door. If you've taken a long route and lost a lot of time, grab the path now and begin filling your bank account.

Sounds good so far? Before I dive deeper into the concepts of money-saving, what I want you to do right now, is start building up a money-saving mindset. This is extremely important and will help you during your early days of saving money.

So, what exactly do you have to do?

Building Up A Mindset For Saving Money

The first step? Start imagining.

For many people, the concept of financial independence and never having to worry about money is a dream. They want to achieve, and many of them actually strive to get there.

Imagine having this kind of life: you wake up in the morning, not in a rush to get ready for work and beat the morning traffic, but to a luxurious and relaxed breakfast with your family. You read the papers for an hour, then spend the next hour on the phone talking with your managers and financial advisors, getting updates on how your investments are doing. Once you confirm that everything is going well with your investments, you can join some friends for a game of golf. In the afternoon, you would spend time working on one of your hobbies—whether painting, playing an instrument, writing, you name it. Every summer, you get to take your family on vacation to a tropical island and to spend your days sipping on mojitos and piña coladas. At the same time, you watch your kids frolicking on the white and sandy beaches. All this without having to worry about money or limited leave days. Wouldn't you want such a life?

The truth is that most people would want that life, though many do struggle to build such a life. However, for many of them, this kind of life doesn't go beyond just being a dream. This is because you can't just get to wake up one morning and start living such a life. Such a life, a life of financial freedom, results from careful planning and deliberate actions. Those who live such lifestyles got there through lots of hard work and smart moves.

However, the good thing is that if you know what to do and if you actually commit doing it, you can attain financial independence. Below are some of the things you need to do if you want to be financially independent.

1. Start Spending Less Than You Earn

When I used to work as a consultant for a top management firm, I was earning quite a high salary, yet I was in debt and living paycheck to paycheck. The problem was not how much I was earning; it was how I used to spend my money. Eager to live a flashy lifestyle, I used to purchase things I couldn't afford, which is why I found myself in debt.

If you want to attain financial independence, the first thing you need to do is to start living within your means, which means spending less than you earn. Without doing this, it is impossible to become financially independent, regardless of how much money you make. Ideally, you should aim for a situation where you get to save at least 30% of your monthly income.

Therefore, if you have been spending beyond your income, then find ways to start spending less first, which brings us to my second point.

2. Cut Down Your Expenses

If you have been spending more than you earn, the best way to start spending less is to cut down your expenses. You also have the option of increasing your income, though this is a lot harder than cutting expenses. Even if you have been spending less than you earn, cutting down on your expenses will allow you to save more money.

To cut down your expenses, you need to know where your money goes. Start by tracking everything you spend your money on every month. Once you have a good idea of where your money is going, go through the list and identify some unnecessary expenses or those you can cut back on. For instance, if you pay money for the gym, but you rarely ever set foot in the gym, you need to cut down on that expense. If you are making car repayments, you could sell the car and find a cheaper one with lower repayments. You may even purchase a cheap used car that you can afford without taking on debt. If you live in an expensive apartment, consider moving to a smaller, less costly apartment. Basically, if you have any unnecessary expenses, get rid of them and find ways to reduce the cost of your necessary expenses.

3. Pay Off Debt

If you want to become financially independent, you also need to pay off any debt you owe. There are several reasons why you should focus on repaying your debt.

First, having lots of debt means that you will have equally high payments to make every month. Once you reduce your debt, your loan payments will also go down, leaving you with more money to save. For instance, if you were making debt payments worth $600 each month, lowering your debt might lower your monthly loan payments to $300. This results in an extra $300 in your account every month, or about $3,600 every year. If you clear all of your debt, you will have an extra $600 each month, or about $7200 every year.

The second reason, why you need to pay off your debt is that debt costs you money. The longer you take before clearing your debt, the more money you will end up paying in interest charges. This is money you could have saved by clearing your debt.

Finally, paying off your debt will lead to an improvement in various indicators of your financial health, such as your credit score. This will, in turn, make it easier for you to get access to financing in the future.

For instance, if you decide to invest in real estate but don't have enough money saved up to pay for a property, a good credit score will make it easier for you to get financing.

4. Find Ways to Increase Your Income

Now, that you have cut down your expenses and paid off your debt, start thinking of ways to increase your income. You can do this by asking for a pay raise at your current job, starting a side hustle, finding a new job that pays better, renting off a room in your house on Airbnb, selling off stuff in your home that you no longer need, and so on.

One thing to note here is that while you are increasing your income, you need to watch out to ensure you don't increase your expenses. The reason you are increasing your income is to allow yourself to save even more money every month.

5. Make the Right Investments

All the previous steps have been focused on ensuring that you have more money to save each month. However, you are not saving this money just for the sake of it. The money you are saving will act as the foundation that will allow you to attain financial freedom, which will only happen if you invest the money you have saved up. Investing is what allows your money to multiply.

There are several ways through which you can invest your money, such as purchasing stocks, bonds, trading in commodities, peer-to-peer lending, putting your money into mutual funds, high yield savings accounts, and so on.

However, one of the best ways of investing your money is putting it into **Real Estate**. This is the quickest way of growing you towards financial independence.

Like, I mentioned in the introduction to this book, most millionaires have investments in the real estate sector, and a lot of them attribute their wealth to returns from their real estate investments.

Some of the richest people in the world who made their wealth through investing in real estate includes Chinese billionaire Wang Jianlin, who has an empire valued at over $28.7 billion; Lee Shau Kee, whose empire is more than $21.5 billion; German billionaire Michael Otto, who is worth $15.4 billion; and American billionaires Donald Bren, Stephen Ross, and David Lichtenstein, who have a net worth of $15.1 billion, $12 billion, and $1.45 billion respectively.

6. Compute Living Budget and Expenses

Finally, to become financially independent, you will need to compute how much money you need to live comfortably. This amount will differ from person to person, depending on lifestyle, preferences, living arrangements, family size, and so on.

Assuming you have already made an investment in a rental property, you will want to get in a position where your rental income can cater to all your living expenses. From there, as you expand your portfolio, your investments will generate more money than you need to survive, which will act as your launchpad on your journey to financial freedom.

Following and incorporating above 6 steps into your life and mindset will help you preparing coming days. But, like I said, these are just the start.

CHAPTER TWO

Having The Clear Saving Goals

Knowing how to save your money is one thing, but the thing that will motivate you the most to save your money is to essentially have a very well-thought-out savings goal. Many people think of setting goals as just simply having a target and working towards reaching that target. While that is true in most cases, a Savings Goal is slightly different and requires a bit more effort than that.

The first thing that you should understand regarding a Savings Goal is why you should even have a savings goal?

Why Should You Bother With Savings Goal?

As humans, we always need something to look forward to, and with our busy schedules, we need time to plan things out. Many people who are bad at saving may not have set any goals at all, or they may have set goals that were entirely unrealistic to achieve.

Setting financial targets will help you keep on track. You know you want to go on that cruise or take the family to Disney World, but you may not understand what it takes to get what you want if you don't prepare ahead of time.

Setting objectives allows you to break down the saving process and make it simpler and more convenient. Setting goals allows you to give yourself more time to save. Many people are still very impulsive with their money, and as a result, they seldom have enough cash on hand to buy the things they really desire.

A key part of setting up your Savings Goals is to understand your "Needs" and your "Wants."

You can set targets under both situations, but to really know what you can and cannot afford, it is a good idea to first set savings goals to decide whether you can accomplish the things you want while still having enough money to afford what you need.

How might a Savings Goal look like?

Some Quick Examples Of Savings Goal

1. **Saving For College**

Put money aside for education – A $20,000 college education in 1977 will cost $302,434 in 2020, but a degree still pays off. According to the United States Department of Education, college graduates with a bachelor's degree receives 66 percent more than those with just a high school diploma. The gap in earnings over a lifetime is at least $1 million. According to the Georgetown report, in 2020, an estimated 35% of all job openings required at least a bachelor's degree and 30% requiring an associate degree or another sort of education.

2. **Saving For A New Home**

Save for a down payment on a house – For most people, this is their largest purchase and investment. The greater the down payment, the greater the freedom and flexibility offered during the loan's term. A 20% down payment is the industry norm for a decent mortgage. Remember that owning a home is a much better investment than renting.

3. **Saving To Pay Off Student Loans**

Paying off the student loans – For millions of Americans, crippling student loan debt is a relentless burden. The estimated student loan debt for 44.7 million Americans is $32,731. If you want to pay off your student loans quicker, consider refinancing at a lower interest rate. However, if you have governmental loans and want this choice, be cautious about refinancing with private lenders. Certain safeguards, such as income-based repayment, deferment, and forbearance, are no longer available after refinancing with a private lender.

4. Saving To Start A Business

Creating a company – Creating a business is a difficult but eventually rewarding undertaking. Who doesn't want to be in charge? When starting a company, you must develop a business plan, raise seed capital, and adhere to a monthly budget that is within your means. Starting a company is intended to make money, not to lose it. Maintain your fortitude!

5. Saving For Retirement

Save for retirement – For some Americans, the idea of delayed gratification remains elusive. "Everything around us is a push to buy, a push to consume," said Annamaria Lusardi, a distinguished professor of business at George Washington University. "We need to make a saving, particularly retirement saving, as exciting as consumption. And it's exciting because it offers us the ability to achieve our long-term goals. People simply ought to see things in that light." Set money aside each month to build your retirement portfolio. You'll be glad you did it later.

6. Saving To Pay Off Your Card Debt

Pay off credit card debt – According to Wohlwend, this quality should be at the top of the list for anyone serious about setting financial standards. "The interest charges (on credit card accounts) consume too much of the cash flow that could be used for other purposes," Wohlwend said. "Once you've paid them off, you should be careful not to use the credit card too much. People can make bad choices because the system as a whole encourages them to do so. When you get swept up in the world, you don't even realize what's going on until you add it all up. It's as if they're saying, "Oh my gosh, I'm $150,000 in debt!" If you're having trouble doing it on your own, consider credit consolidation through a reputable charitable credit counseling agency.

Now, the question is, how are you going to set up your goal and start saving for it? Well, we will go into the concept of Long-Term and Short-Term financial goals in a moment. Still, before that, you should know that there is a very basic rule of thumb when it comes to setting up a Savings Goal.

The Rule Of Thumb Of Having A Financial Goal

1. Clearly Lay Out Your Goals

To begin, you must identify your savings objectives.

- Make it specific! Make a list of two or three savings targets you have.
- Set a deadline for each goal. This allows you to differentiate between short-term goals, which can be accomplished in one to three years, and long-term goals, which require more than three years.
- Also, write down why you want to accomplish each of your objectives. Make yourself as transparent as possible. The "why" keeps you going.

2. Pick Your Saving Methods

Following that, you must choose your saving methods. This can be determined by the timeline for your goals as well as your risk perception. A high-yield savings account serves the good purpose, where you need money faster or want to have quicker access to them for the short-term objectives. It allows you to gain higher interest while still providing you with immediate access to your funds with no taxes or fees.

Longer-term objectives can necessitate the use of higher-yielding savings vehicles, such as investment accounts. Alternatively, you could have a savings account that is managed by your employer. Always consider when you will need the money, and be aware of any taxes, fees, or penalties associated with the account you choose. A separate account for each target is often supported by a large body of proof.

3. Understand Your Budget

It's now time to calculate how much money you will save each month. This entails making a budget.

- Examine your expenses over the last few months.
- Determine how much money you'll need to cover your essential expenses.
- Be sure to plan your monthly bills, groceries, mortgage payments, and other costs that you must pay to get through the month.
- Take a look at what's left.

This is where you can find savings opportunities. Determine what you spend money on per month that isn't important. What would you slash? How much you cut depends on how much money you want to save and how hard you want to work to achieve your objectives. It can be beneficial to keep some **"fun money"** in your budget rather than allocating every spare dollar to savings. This allows you to remain on track and not feel so constrained. Determine how much you can safely remove while keeping in mind that you do not have to eliminate all discretionary expenditures. Simply cutting back will help you save money.

4. Divide Your Accumulated Savings Between Goals

You must now split your savings among your objectives.

- Determine how many months you have to achieve each target using the goal completion dates you wrote down.
- Divide the total amount you need to save by the number of months you want to achieve each target. This will tell you how much money you need to save per month to achieve your target.
- Divide your monthly savings between each target based on how much you have to save for each.
- If you don't have enough money to cover all of your goals, you'll have to prioritize. Some objectives, such as an emergency fund, are time-sensitive. Other priorities, such as retirement savings, can be prioritized to benefit from an employer match.
- Consider your written goals and how achieving each one represents your ideals and affects your overall financial image.
- Fully finance your top priority and apply any remaining savings to your next priority. If you've completed your top target, you can apply your savings to your next goal, allowing you to complete it faster.

5. Keep Track Of Your Progress

You'll want to keep track of your progress now that you've developed a savings plan. If you use a pen and paper, Excel, or an app, keep track of your monthly savings for each target. This will help you stay inspired!

6. **Repeat**

Setting savings targets is a continuous process. You'll have to repeat the process as things change. And when you reach your short-term savings targets, you'll want to restart the process and set new goals or contribute more to your current ones. Whatever you do, remember to celebrate your victories and successes along the way! Recognizing your accomplishments will help you stay motivated in the future.

Following the steps above will help you stay on track when pursuing a Savings a Goal and make sure you are making progress.

Let me now walk you through what long-term and short goals actually are with that out of the way.

Understanding The Three Different Types Of Financial Goals

Setting financial targets is one of the most effective ways to take financial responsibility in your life. By taking the time to understand your life goals and the amount of savings needed to achieve them, you are taking a meaningful step toward eventually realizing them.

You won't know where you're going if you don't have goals, and you won't be able to get there. If you have ambitions, it is impossible that you can achieve them unless you devise a financial strategy to fulfill those goals.

There are three different types of goals that you should consider while planning out your financial outline.

1. Short Term.
2. Mid Term.
3. Long Term.

We will look into each of them one by one, but before that, let me give you a brief definition of what each of them means.

1. Short Term

Your more urgent expenditures are your short-term ambitions. While timelines vary, these are the items you will most likely spend money on over the next few months or years.

Example of a short-term goal:

- An emergency fund.
- Share, insurance, or student loan payments.
- Payments on credit card interest.
- Personal belongings
- Traveling.
- Wedding.
- Minor home maintenance and upgrades.

2. Long Term

Long-term objectives are usually the large-picture costs. These objectives can take years, if not decades, to achieve.

Long-term goals usually require more resources and consistent effort than short-term goals.

- The retirement fund is an example of a long-term target.
- Getting a mortgage paid off.
- Establishing a company.
- Putting money aside for a child's college education.

3. Mid-Term

There is sometimes overlap between the two groups, which may cause confusion. Medium or mid-term targets slip between short-term and long-term objectives and typically take several years to complete. Examples include

- Purchasing a car.
- Putting money aside for a down payment.
- Paying off debts.

You with me so far? Now let's talk about how you can set up these goals.

Setting Up the Short-Term Finance Goal

Setting short-term financial targets will provide you with the confidence boost and foundational skills you need to achieve longer-term goals. These initial steps are relatively simple to complete.

While you can't make a million dollars appear in your savings account right now, you can sit down and make a budget in a few hours. Many people can save a good emergency fund in a year. Here are some main short-term financial goals that will benefit you immediately and help you tracking your longer-term goals.

1. **Set Up An Budget**

"You can't know where you're going unless you know exactly where you are right now. This entails creating a budget," says Lauren Zangardi Haynes, a fiduciary and fee-only financial planner with Spark Financial Advisors in Richmond and Williamsburg, Virginia. "You can be surprised by how much money slips through the cracks each month."

Using a free budgeting app like **Mint** is a simple way to keep track of your spending. It will consolidate details from all of your accounts and allow you to categorize each expense. You can also make a budget with the old-fashioned way by going through your bank statements and bills from the previous few months and categorizing costs on paper or in a spreadsheet.

You can discover that ordering Seamless every work-from-home day (or spending that much on lunches with coworkers if you're back in the office) costs you $315 a month, based on $15 per meal for 21 workdays. You could discover that you're spending an additional $100 per weekend on date nights with your significant other.

If you've seen how you're investing your money, you can use the knowledge to make informed choices on where you want your money to go in the future. Is the pleasure and comfort of dining out worth $715 a month to you? If so, that's fantastic—as long as you can afford it. If not, you've just found a simple way to save money every month.

You should look for ways to spend less money while dining out, replace some restaurant/takeout meals with homemade ones, or do both.

2. **Create An Emergency Fund**

An emergency fund is a money set aside expressly for unanticipated expenditures. Aiming for $500 to $1,000 is a reasonable place to start. If you've reached that target, you'll want to increase it so that your emergency fund can cover bigger financial problems, such as unemployment. If you didn't have an emergency fund before the COVID-19 pandemic, you really wished you did now. If you had one, you might have depleted it and need to replenish it.

Ilene Davis, a certified financial planner (CFP) with Financial Independence Services in Cocoa, Fla., suggests saving at least three months' worth of expenses to meet your financial commitments and basic needs, but preferably six months' worth especially if you are married and work for the same company as your spouse or live in an area with little job opportunities. She claims that having at least one area in your budget to cut back can help finance your emergency savings.

According to Kevin Gallegos, vice president of sales and Phoenix operations for Freedom Financial Network, decluttering and arranging are two more ways to save for an emergency, an online financial service for consumer debt settlement, mortgage shopping, and personal loans. You can supplement your income by selling unwanted products on eBay or Craigslist or hosting a yard sale. Consider converting a hobby into part-time work so you can put the extra money into retirement.

Opening a savings account and setting up an automatic transfer for the sum you've decided you will save per month (using your budget) before reaching your emergency fund target is recommended by Zangardi Haynes.

"If you receive a bonus, tax refund, or even an 'extra' monthly paycheck—which occurs twice a year if you are paid biweekly—save the money as soon as it enters your bank account. If you wait until the end of the month to move the money, the chances are it will be spent rather than saved" , she says.

Although you most likely have other savings objectives, such as planning for retirement, establishing an emergency fund should be your top priority. The savings account provides the financial security needed to achieve your other objectives.

3. Clear Off Your Credit Card Debt

Experts differ about whether it is better to pay off credit card debt or save for an emergency. Some argue that even though you have credit card debt, you should set aside money for an emergency fund because any unforeseen cost would push you deeper into debt. Others argue that you can pay off credit card debt first because the interest is so high that it makes any other financial target far more difficult to achieve. Choose the philosophy that makes the most sense to you, or combine elements of both.

Davis suggests listing all of your debts by interest rate, from lowest to highest, and paying only the minimum on all but the highest-rate debt as a method for paying off credit card debt. Use any spare money you have to make extra payments on your highest-interest card.

The debt avalanche is the process, Davis explains. Another approach to thinking of is the debt snowball. The snowball approach requires you to pay off your debts in the order of smallest to highest, regardless of interest rate. The theory is that the feeling of satisfaction you get from paying off the smallest debt would motivate you to take on the next-smallest debt and so on until you're debt-free.

According to Gallegos, debt negotiation or mediation is an option for those who have $10,000 or more in unsecured debt (such as credit card debt) and cannot make the regular minimum payments. Companies that provide these services are supervised by the Federal Trade Commission and operate on behalf of the borrower to reduce debt by up to 50% in exchange for a charge, which is usually a percentage of the overall debt or a percentage of the amount of debt reduction, which the consumer can only pay after a satisfactory agreement. Gallegos claims that this method will help consumers get out of debt in two to four years. The disadvantages of debt settlement are that it can harm your credit score and that creditors can take legal action against customers for unpaid accounts.

Setting Up Mid-Term Finance Goal

If you've built a budget, formed an emergency fund, and paid off your credit card debt—or at least made a significant dent in those three short-term goals—the time it's to focus on your long-term financial goals. These objectives would serve as a connection between your short- and long-term financial objectives.

1. Consider Getting A Life Insurance

Do you have a spouse or children who depend on your earnings? If this is the case then you would need life insurance to protect them if you die prematurely. Term life insurance is the simplest and least expensive form of life insurance. It can meet the majority of people's insurance needs. An insurance broker will assist you in obtaining the best deal on a policy. Most term life insurance policies require medical underwriting. Unless you are terminally ill, you should be able to find at least one insurer willing to give you a policy.

Gallegos also recommends that you have **disability insurance** to cover your income when employed. “The majority of employers have this coverage,” he says. “If they don't, people will get it before retirement age.”

Suppose you become severely ill or disabled to the point that you cannot function. In that case, **disability insurance** can cover a portion of your wages. It can offer a greater advantage than Social Security disability benefits, allowing you (and your family, if you have one) to live more comfortably than you would otherwise be able to if you lose your ability to work. There will be a period between when you become unable to function and when the insurance plans begin to payout, which is why maintaining an emergency fund is critical.

2. Pay Off Any Pending Student Loan

Student loans are a significant burden on many people's monthly budgets. Lowering or eliminating such payments will free up cash, making it easier to save for retirement and reach other goals. Refinancing into a new loan at a lower interest rate is one tactic that will help you pay off your student loans. However, suppose you refinance federal student loans through a private lender. In that case, you can sacrifice some of the incentives associated with federal student loans, such as income-based repayment, deferment, and forbearance, which may help you get into financial difficulties.

Suppose you have several student loans and cannot benefit from consolidation or refinancing. In that case, the debt avalanche or debt snowball strategies described above will help you pay them off faster.

3. **Bring Your Dreams Into The Picture**

Mid-term objectives may also involve purchasing a first home or, later, a holiday home. Perhaps you already own a home and want to improve it with a major renovation—or you want to start saving for a larger house. Mid-term goals include saving for college for your children or grandchildren, as well as saving for when you do have children.

If you've decided on one or more of these objectives, start calculating how much money you'll need to save to make a difference in meeting them. The first step in reaching the desired future is to visualize it.

Setting Up Long-Term Finance Goal

Many people's primary long-term financial target is to save enough money to retire. The general rule of thumb is to invest 10% to 15% of each paycheck in a tax-advantaged retirement account such as a 401(k) or 403(b), whether you have access to one, or a conventional IRA or Roth IRA. However, to ensure that you are really saving enough, you must first determine how much money you will need to retire.

Understand Your Retirement Requirement

Oscar Vives Ortiz, a CPA financial planner with PNC Wealth Management in the Tampa Bay/St. Petersburg area says you can estimate your retirement readiness with a simple back-of-the-envelope calculation:

1. Calculate your expected annual living expenses in retirement. When you first started working on your short-term financial goals, the budget you made would give you an idea of how much money you'll need. In retirement, you will need to budget for higher healthcare costs.
2. Subtract the amount of money you will get, including Social Security, insurance, and retirement benefits. This would leave you with the sum that your investment portfolio would cover.
3. Calculate how much money you'll need for retirement based on your expected retirement date. Calculate this based on what you actually have and how much you save per year. You can do the math with an online retirement calculator. Suppose that 4 percent or less of this balance covers the remaining costs that your combined Social Security and pensions do not cover at the time of retirement. In that case, you are on track to retire.

If you started with a $1 million portfolio and removed $40,000 in year one (4 percent of $1 million), then raised the withdrawal by the rate of inflation per year ($40,000 plus 2% in year two, or $40,800; $40,800 plus 2% in year three, or $41,616, and so on), you would have made it through every 30-year retirement without running out of money.

"This is why, when addressing retirement, 4 percent is sometimes used as a rule of thumb," Vives Ortiz says.

"In most cases, using 4 percent results in more revenue at the end of 30 years, but in the worst-case scenario, you would have run out of money in year 30," he says. "The only word of warning here is that just because 4% has survived any situation in history does not guarantee that it will continue to do so in the future."

The next thing that naturally comes to mind is how you prioritize between long-term and short savings?

Prioritizing Long Term And Short Term Goals

Setting long-term goals first is a smart way to prioritize goals. This will give you a sense of mission and direction. Once you've defined and established your long-term objective, you can begin setting short-term goals that contribute to your long-term goals. Short-term goals are stepping blocks to achieving long-term goals. They are simpler and easier to achieve than long-term objectives. Achieving short-term goals provides a feeling of achievement that will inspire you to achieve your long-term goals. Setting and working on these short-term goals, as well as achieving them, gives you momentum toward achieving your long-term plans.

Until you begin setting goals, take a few moments to make a list of the things that are most important to you. Writing down your goals will help you to remember them.

Most importantly, remember to reward yourself if you accomplish any of your objectives.

To sum things up, You are unlikely to make flawless, linear progress toward all of your goals. Still, the important thing is to be consistent rather than perfect. If you have an unforeseen car repair or medical bill one month and cannot contribute to your emergency fund but must withdraw funds from it, don't be too hard on yourself; that's why the fund exists. Simply get back on track as soon as possible. The same holds true if you lose your work or become ill. You'll need to devise a new strategy to get through that tough era. You may not be able to pay down debt or save for retirement during that time, but you can revive your original plan—or maybe an updated version—once you've made it through.

That is the beauty of annual financial planning: you can revisit and refine your goals while still tracking your progress against them through life's ups and downs. Throughout the process, you will discover that both the small things you do on a regular and monthly basis and the big things you do every year and over decades will help you reach your financial goals.

CHAPTER THREE

Awesome Strategies to Start Saving Money

Now you know, why it is important to save money and effectively make a plan for savings. This chapter will discuss some awesome strategies that you can use to save those precious pennies.

It takes time to achieve long-term financial success. It takes time for investments to grow, compound interest to work, and career changes to pay off.

The issue is that when you are dealing with a financial crisis, you normally don't have a lot of time. The bills keep piling up. Debtors are contacting you. A high cost has just been slapped on your lap. You must make financial improvements as soon as possible.

Frugality is important in these cases. What you need are effective ways to reduce your expenses without significantly reducing your quality of life. Choose one of these tips and apply it to your life, where it makes sense to reduce your expenses without affecting the necessities. So, that you have enough money to pay your bills and stabilize your financial condition.

Planning Is Key

As mentioned earlier, saving money requires a plan. For you to do it responsibly, there are steps that you need to consider. You should be aware that you can't wake up one morning and start saving without a plan and expect that you will do it for a long haul. You will find that approaching a savings plan on a whim and really not thinking it through will cause you to lose your focus. The first thing that you have to learn when planning to start saving is how to do it responsibly.

The following steps provide you with ideas on how you can start to save money responsibly

1. **Start by paying yourself**

In most cases, people invest their money as soon as they receive it. They always waste their money before they even have it (i.e., they borrow on credit cards, from their family, etc.). They simply take the money and begin spending and accruing debts without giving it much thought. When you get your money, I suggest that you start by paying yourself first. Allow a portion of the funds to be deposited into your account. A savings account, to be specific. You will put another part of your earnings into a savings account if you have one. No matter how tiny the deposits in your savings account are, they will add value and give you the trust to make additional deposits in the coming weeks and months. You may also set up an automatic deposit to your savings account. This should be a separate account from your regular checking account. Contact the company's payroll department or the individual in charge of cutting payroll checks. Automatic deposits are normally possible to set up. With a little effort, you can set up a portion of your income to be routed to different savings accounts.

2. **Avoid accumulating debt**

There are some debts that you can't avoid in life. Such debts include loans to buy yourself a car or a house for your family. Only wealthy families can buy houses and cars by paying for the cash in one lump sum. Most people need to pay a deposit on a home or car and then repay the remaining amount in installments over some time. If you can get out of debt, try to pay off your debts as soon as possible. Have you ever thought of only paying for things with the cash you already have as opposed to buying things on credit and then repaying the loan with interest?

You should always avoid other debts like credit card charges when purchasing things that you can do without. These purchases are unnecessary— you know, the ones I am talking about. That new flat screen television that you just bought on credit when you had just bought a flat-screen television just two years ago. Or that seven-day cruise that you charged because you wanted to impress your girlfriend.

Now is the time that you must have a clear understanding of what "spending responsibly" means. It starts by being responsible and disciplined when managing your money. The next step in your savings plan will involve finding ways in which you can reduce your expenses.

The following steps will give you a clear idea of how you can cut your expenses and still live a very fairly respectable lifestyle

1. **Removing luxuries from your planned budget**

When planning on how to spend for a particular month, you will find that there are some things that you end up buying that you can do without. This is a key factor in success when trying to save money - cutting down or cutting out luxury purchases.

You might think that it is very difficult to live without cable or satellite TV, a huge gas-guzzling car, and so on. But it is very easy to live without these things. In fact, you can live a very nice and fulfilled life with just the essentials. You might not even imagine how good your life will be.

Consider thinking how much you might save when you

- Unsubscribe from internet packages and television packages which are optional and really are luxuries.
- Buy your home furniture and clothing from thrift stores
- Take unused electronic devices in your household and sell them.
- Buy a car that is easy to maintain and economical in fuel as opposed to an unnecessary luxury vehicle such as an SUV, which is horrible on gas mileage per gallon.
- Switch to a more economical phone service plan once implemented, you will find that you will have saved a lot of money at the end of the day. Luxurious things take a very large portion of monthly income. Cutting them out is a sure way of accumulating extra money for savings. These things seem obvious, but many people spend their money month in and month out and then wonder — 'Where did all my money go?'

2. Find a cheaper house

Most people spend 40% of their monthly income on rent or mortgage payments. Others have to pay even more. If you're having trouble budgeting your finances, consider the type of house you're currently living in. If you are a renter, consider contacting your landlord and negotiating a lower rent before looking for cheaper accommodation. Many homeowners would rather have a current tenant who pays marginally less than a brand new, inexperienced tenant who can pay more.

The landlord is accustomed to the current tenant (you) to understand that you pay your rent on time, are quiet and ethical, and generally behave well. A new occupant brings with him or her unknown actions and risks.

It can be difficult for the landlord to predict whether the new occupant would positively or negatively. As a result, you might have the ability to get a great deal on rent.

Otherwise, you might consider moving to a less expensive area and looking for a place to live where the rent is less expensive.

3. Reduce your food costs

Most people want to spend frivolously on food. They spend their money in expensive restaurants and buy foods from specialty grocers. Many of these types of foods are associated with unhealthy eating and generally unhealthy lifestyles. Why not consider buying your food from less expensive grocery stores, buying in bulk, and eating out at less expensive restaurants? As long as the quality is good or least adequate, this is the best way to go. Another option is to not eat out at all, or only on special occasions. At the end of the day, you will discover how much money you are wasting by dining out, and you will save quite a bit of money.

Instead of buying processed and prepared foods, consider buying more produce items and fresh foods from the lower-cost grocery stores, enabling you to eat healthier.

Always consider sales when shopping and incorporate using coupons every week from both restaurants and local grocery stores. Furthermore, it is usually more economical to cook at home rather than eating out at restaurants.

4. Consider reducing your energy usage

It is relatively simple to reduce the energy intake, and large amounts of energy can be avoided. Small changes, such as turning off the lights before you leave the building, can make a big difference. If you find it difficult to remember to do this, consider leaving a post-it note on your front door when you leave the house to serve as a reminder. Another option is to stop using air conditioning and heating when they aren't absolutely required. Why not try opening your windows or using a slew of fans? If you want to keep it inside, you can use a space heater, cover yourself in a blanket, or simply dress in extra layers.

Investing in solar panels is another way to cut electricity costs. Solar panels can generate natural electricity from a variety of light frequencies. This move would be a bit drastic, and the initial costs could be significant.

However, the long-term costs are lower because you can pay no other energy costs — thereby eliminating the premium.

You gain prestige by owning a vehicle, and by owning a car, you blend in with the average customer. Friends and family members show you "respect." But have you ever considered any of the costs associated with driving and maintaining a vehicle? Depending on how many cars you have in the family and how much they are driven, the cost of gasoline alone could cost you hundreds of dollars a month. You will also have to pay recurring maintenance and registration fees.

Why not to reserve your car for important occasions such as weekend celebrations, appointments, and family outings? This will allow you to use less costly or free transportation alternatives for your day-to-day business operations. You could end up spending a lot of money at the end of the month if you start using public transit, depending on where you live.

Buses, subways, and metro trains are available in most major cities. Carpooling is another way to save money. Perhaps you have a coworker who lives in the neighborhood. It's not a bad idea to inquire around. Fuel savings will add up over several months and years.

5. Having cheap and/or free fun

Ever heard of fun for cheap? Well, this doesn't mean that you should stop having fun. It means cutting the frivolous and luxurious partying type of fun like clubbing and taking lavish vacations. You could replace that type of fun with recreational activities, which are more affordable. Adopt healthy and cheaper social habits.

Why not consider reading books rather than purchasing video games and going to the movies? Books are very cheap. You can get them from the used book store or perhaps on eBay or Amazon. You can also enjoy your social life with activities like watching movies from less expensive theaters or going at discounted times (matinee), playing board games, hiking and spending time at the park

Keeping up with community events is another way of enjoying your free time and is a sure way of spending less money.

6. Avoiding expensive addictions

Certain bad habits can surely hinder your money-saving efforts. Without outside help and support, you might find it hard to overcome such addictions. Remember that some addictions are very dangerous to your health in the long run. Avoiding these addictions is the key.

Smoking is one such addiction. Quitting smoking will benefit your wallet and your health. These days, most people should have an understanding of the negative effects of smoking. They include high blood pressure, risk of stroke, heart disease, and lung cancer - just to mention a few.

Excessive alcohol consumption is another habit that you should avoid. Drinking two or three beers with friends is usually okay and considered socially acceptable. However, drinking heavily regularly is harmful to your health in addition to interfering with your money-saving efforts. Remember that your family might also have to bear the financial burden of treating you in the case of your alcohol addiction.

Other bad habits that you should avoid include using drugs such as marijuana, cocaine, and heroin. I think this advice is quite obvious, but sometimes when one is amid addiction, they may not realize the damage being done to their health and the financial implications. Remember that these drugs are very expensive. Once you are addicted, your life can spiral downhill, and the financial implications can be severe. Even worse, the damage to your health could be irreparable.

You now have a general idea of how to save and spend responsibly. You also have an idea of ways to save a lot of money by cutting out unnecessary and luxurious expenses. The next step is to learn to spend your money intelligently. Remember that you don't want to deny yourself a social life and other essential things that life offers.

The following steps will provide you with proven ways through which you can consider spending intelligently

1. Buying only the absolute essentials

Some things in life are absolute essentials for you to live. These are the things that you will purchase after you have allocated a specific amount aside for savings for that month. These items include food, clothing, housing, electricity, and water. They should be on priority list each month within your allocated funds.

Before spending your money on other things, purchase the absolute essentials first. Keep in mind that you should not use these essentials as excuses for your heavy spending. Be mindful of how to cut down these expenditures. Therefore, spend wisely to ensure that you use the least amount possible while keeping a moderate comfort level.

2. Savings for emergencies

You should have a separate savings account where you will be saving emergency funds. This account is so important ensuring you to rely on these funds in reduced incomes and unemployment. Remember that life has to continue if you lose your income. You need to stay on top of your monthly expenses with housing, food, utilities, and all necessary expenses to make ends meet.

Saving for emergency situations is a sure way of reducing the stress associated with losing your job and not knowing how long it may be until your next paycheck. Remember to determine the number of your emergency funds to cover the expenses that you incur per month.

3. Paying off your debts

You have to take this mover after you've bought your monthly necessities and saved for an emergency fund. You will severely jeopardize your money-saving efforts, if you do not keep your debts on the top of the strategies. Maintaining debt and failing to pay it off quickly would only make the debt more difficult to repay in the future. It will accrue interest, which you do not want to do. The easiest way is to pay off your debts a little at a time, faithfully and on time, each and every month. Pay more than the monthly minimums if possible to pay them off faster.

Pay off these loans as quickly as possible. I suggest beginning with the debts with the highest interest rates and then progressing to debts with lower or even no interest rates. If you have debts from several lenders, you should think about consolidating them. This can be accomplished by borrowing from a single lender and consolidating all debts into a single loan with the lowest interest rate. This is referred to as **Debt consolidation.**

4. Putting some money in your savings account

This step occurs after you have paid all your debts for a given month. It will depend on the plan you mapped out for the month and paying all your debts first. Remember that you are now saving this portion of funds in a savings account (opposed to the emergency funds). The savings in this category can be used for any other life necessity such as buying yourself a car, a house, paying your children's school tuition, and so on. You can also make bigger purchases with these funds such as a new computer, flat-screen television, a mini-vacation, etc. Devoting 15% of your income for the purpose of saving is a smart objective. Many financial experts agree that saving money each month by budgeting and allocating funds to a saving account is highly recommended — but many times difficult for some to implement. If you can discipline yourself to stick to this plan, you will be ahead of the curve in a money crunch. One way to ensure these funds are allocated to the savings account is to set up an automatic withdrawal from your paychecks to automate a certain percentage into your savings account. Again, this can be done through your payroll department at work

5. **Consider spending on smart non-essential purchases**

Now, what do we mean by this? This step should follow after you have allocated a certain percentage of your monthly income for savings. Obviously, you have a very long list of essentials, but some are more important than others. For instance, you can consider purchasing an ergonomic chair for your office or living room. Using this chair in your office is not essential for you to continue working in your office.

However, in the long haul, you will increase your productivity and reduce the chances of experiencing back pain. Smart non-essential purchases allow you to increase your earning potential, improve your life and improve your productivity. It is another way of saving money.

6. **Finally, you can now purchase luxury items**

You should always remember that saving money doesn't mean that you should be too lean and hard on yourself. After you have spent on the essentials, saved for emergencies, paid your monthly debts, saved some money, and spent on your smart essentials, you can now use the remaining funds for luxury purchases. Luxury items really mean you can now spend any leftover disposable income, which means you can buy whatever you want. However, when talking about luxuries, this is a very broad category of spending. You can consider purchasing a new car, paying for cable television, going on vacations, or going out to expensive restaurants. You have earned it based on your saving habits and discipline throughout the month. Always be reasonable when spending on luxury items. They should only come after everything else on your budget list..

Organize Yourself To Save

1. **Be Honest To Yourself**

Let's be real. Ignorance is bliss, but it's still ignorance. It's time for you to understand the full picture. If you have debt, you need to understand how you got there because you can't fix it if you can't see what it is. It's time.

- Stop making excuses.
- Stop avoiding it.
- Face any fears you might have.
- The best place to start? Let's get your financial documents and accounts organized.

My strategy to clean up the physical paper documents I had lying around (think tax forms, bills, notices from your bank, etc.) was to collect everything of importance into a little cardboard container from IKEA — you can get fancier if you'd like! and keep that box visible on a bookshelf. Weekly, I would open it up and see if I could remove or organize it further. My goal was to get the unruly pile of financial rubble strewn across my apartment into one organized and streamlined system. This worked for me;

- You may prefer a desk organizer or drawer system. Digitally, I made sure I had logins for all of my accounts and bills. There were some I didn't even realize I could access online, like my Health Savings Account (HSA) and all those pesky old retirement accounts.
- Save the login information for these sites in one ultra-secure place (I recommend LastPass).

- While you're getting organized, log into all of these accounts at least weekly to get familiar and build a good habit of checking on them.
- Getting organized both physically and digitally will help you see the full picture.

It's that simple. I'm not asking you to start making changes yet; just take the baby step to sit down and be transparent with yourself. But also... breathe. Remember, this is your year-long commitment.

2. Review An Year Of Your Spending

Now is the time for a deep dive. Look at the last month of bank statements and find your top three financial offenders. Mine were shopping on Amazon, eating lunch out (#JimmyJohnsForLife), and overspending at the fancy boutique grocery store a block from work.

Now, create a quick plan to reduce those expenses. Speaking for myself:

- First, remove my credit card information from Amazon.
- Second, commit to bringing my own lunch for one month.
- Third, just don't enter that fancy grocery store!

Every time I'm tempted to place a new Amazon order, I remember the eye-opening and horrifying moment after I had actually looked at my expenses and realized I casually racked up over $2,000 from shopping there in 2017. You might find that this step alone will help that voice in your head speak up in moments of financial weakness!

Next, compare your monthly spending across the year. Are there any seasonal offenders? One big culprit might be patios with drink specials in those first warm, sunny days after a long and dreary winter. Or maybe March Madness hits you particularly hard (whether your team is winning or losing, you're spending!).

Find your highest spending month and compare it to your lowest spending month. What are some of the reasons for the variation?What are some of the biggest financial offenders from that high-roller month?

3. **Set Your Goals Straight**

Figure out your **"why"** and work towards your **"how"** by determining the steps to get there. While there's tons of advice out there to help us goal-set, I stick to these three simple keywords: challenging, measurable, and achievable.

Here's a personal example of why and how I developed my goal to write this book:

Why: I want to share everything I learned about personal finance so others might feel empowered to take action.

How: Organize all my notes in a way that sounds coherent and relatable. Share my personal finance journey somewhere, somehow... (YouTube? Blog? Book?)

GOAL: Self-publish an eBook about my personal finance journey by January 31, 2019, so that others might be inspired to jumpstart their journey in 2019 and beyond.

Now it's your turn. Set your goals and, most importantly, stick to them! Remember challenging, measurable, and achievable. If you're stuck, here are some examples (be bold and try 2 or 3!):

Get completely out of credit card debt this year to fully turn my attention towards investing and retirement.

- Open and max out a Roth IRA to start building tax-deferred compounding interest.
- Improve my credit score to 725 within 5 years to be successful in getting Pre-approved for my first home.
- Save $5,000 for an emergency fund because life happens.

Up my 401(k) contribution 1% each month until I hit 12% by year's end so that I can start building wealth that can grow towards my retirement. Try out 8o of the 101 tasks in this book because, dammit, I'm determined to get my finances together in 365 days! Enter your goal(s) on the official Money Resolution checklist at the end of the book!

4. Track Your Accounts And Spending

You must be aware of where your money lives and where it goes; otherwise, this is all a shot in the dark. I link all of my accounts through an online service called Betterment.

Another favorite site is Clarity Money: this rolls up to your spending by categories and is less about amounts and budgeting. It can even help spot opportunities to save by pointing out that Netflix or Spotify subscription, for example.

Some people rave about Mint, and others love YNAB (You Need A Budget). To track spending, some might still prefer an old-school Excel spreadsheet budget or even pen and paper!

Check out all these methods and pick a way that works best for you. The most important thing is that your account balances and spending habits are transparent and accessible.

Another benefit of bringing all your accounts into one place is that you can get a clear picture of your net worth. This gives you context on exactly where you are with your debts and assets (including any investments) and can help you track your progress toward your goals.

Watch next on YouTube: The Money Resolution "Personal Capital Tutorial."

5. Link All Your Bills To One Card

Here's a fast, tangible, and simple way to simplify your financial life right now. I highly advise you to consolidate all of your bills into a single credit card. Personally, I charge all of my regular expenditures to one card and all of my bills to another. Keeping my expenses stable and clear helps me to focus more attention on larger financial goals rather than thinking about whether I have enough money spread across ten accounts to pay my bills. Does this imply that you would cancel all of your other credit cards? Or do you want to get new credit cards? You should not necessarily exercise caution when canceling or opening new credit cards since this can hurt your credit score. Examine which cards have the most value to you (such as points or cashback) and reallocate your bills accordingly.

Make it easy for you to monitor your expenditures, particularly if you're doing it alone. If you prefer to keep a close eye on your cash in and cash out, use a debit card. If you're debt-free and want to play the "points game," use your credit card (plenty more on this later).

6. Change The Bill's Due Date

Yes, you can do this! I recommend having as many due dates moved as possible to coincide with your pay schedule. For example, if you are paid on the 1st and 15th, reserve the first payday for housing and the other payday bills. This helps to make your budget predictable and consistent throughout the month.

First, you need to determine which due dates you want to modify (which you should know after taking action on #4). Consider your rent or mortgage, utilities, credit card, car loan, or student loan first, as those are the heaviest hitters. You can decide how you want to space them out and what will work best for your rate of income and overall plan for financial freedom.

Second, reach out to those companies, creditors, and other institutions to make the changes. Typically, as long as you have a positive payment history and don't make these requests often, it should be a fairly easy process. Most companies will allow you to make these changes online or by phone call. Still, occasionally, creditors will want the change in writing.

For some services like Netflix and Spotify, you can't change your payment due date, but there are many workarounds. For Spotify, cancel your current billing cycle. Give it a couple weeks and if you just can't possibly deal with the ads anymore, resubscribe on the day you want to be billed. Similarly, for Netflix, cancel your subscription and sign up on the day you want to be billed.

There is no fee to change your bill due dates, so take advantage of this hack to balance those monthly bills!

7. Track Your Old Retirement Account If You Have One

When I first started to access my retirement accounts, I didn't think I had more than a couple thousand in my past and current plans. After tracking them all down, getting organized, and upping my contribution, I ended up with roughly $50,000 between my IRAs and 4oi(k)s (all of it invested!). Roll over all those pesky companies 401(k) accounts to one place, be it your current employer's plan if eligible or a personal IRA. It takes time, persistence, a sometimes annoying amount of paperwork, and especially patience. But it's absolutely worth it to have your money in one easy-to-access account that you can easily track, monitor, and tweak if necessary.

Here's a personal anecdote: I recently found a few thousand in a 403(b) from an old employer that had been making a measly 2% return for years (we'll learn more in a later chapter, but that's terrible). Rolling that money into my personal IRA, which gains rh on average, was a huge financial boon.

IMPORTANT:

Do **NOT** be tempted to cash out any of your retirement accounts. The tax implications, penalties, and fees are massive (think 20-25%+)!

Save On Groceries

There are basic necessities in life, and then there are luxuries. When talking about saving money, you might skip buying some luxuries, but what about necessities, particularly grocery. There's no escape! You have to do it, but there are ways of doing it right. We offer you advice on how to control your cost when buying the grocery.

1. **Make lists:**

List down! Lists are my favorite tool for maintaining control over everything. Before you step out of your home for grocery shopping, list down what you already have in hand so that you do not overstock and buy unnecessary things.

Planning is the key to success: When you go grocery shopping, plan ahead on what you need to purchase; if you know what you need for your home (again, a list can be very helpful), you would not be tempted to overspend and buy that extra packet of candy for no reason. Today's shops are built so that if you go shopping without a plan in mind or in hand, you will end up purchasing a lot of extra items that are attractively displayed. Utilize technology to its full potential. You will get the most bang for your buck by taking advantage of the internet. You should look for weekly sales and discounts on different items or shops. You may also set up email reminders for such deals to make your life easier.

Furthermore, rather than basing your meal planning on an episode of Master Chef or your own wish list, focus your meal planning on what's on offer in your grocery store's flyer. Check for the best deals, then schedule menus around those ingredients and what you have on hand, and you'll be surprised to see a much lower grocery bill than you were expecting.

2. Replace the junk with healthy food:

Make healthier food choices! Don't laugh at this one-off. It is great that you can actually save money while you become fit. You tend to avoid high-calorie junk foods if you are choosing your food while on a diet. Colas, chips, biscuits, and other bakery items are all unhealthy and expensive too! Great that you can lose weight and not the money by avoiding these items while buying your grocery.

Look out for the nutritional value and number of servings of each food item that you are buying. For instance, you can substitute cookies with fruit, fresh instead of prepared food, and so on. This will lead to better and healthier food choices for your family.

3. Go homemade:

Use your leftover meals for preparing lunch that you or your family needs to take to your educational or workplace. Again, it will make you healthy, but more importantly, you will save money on your meals. Instead of buying your lunch from the canteen or ordering it from some expensive restaurant, you will be able to open your brown bag and eat the homemade food.

You do not necessarily have to spend a lot of time on lunch preparation, though. Even if you do not have leftovers from dinner, prepare a quick lunch of salad, sandwiches, or fruit. Savings and health, both at the same time, are a definite reward.

4. **Avoid wasting food:**

Don't waste any food. Try to prepare the amount of food in line with your needs. Use leftover, freeze them for the next meal or day, or give a totally new form. So, everybody in the family eats it happily but doesn't let any food go to waste. It is like your money going down the drain.

5. **Keep things organized**

Suppose you have your fridge and pantry organized. In that case, chances are high that you will make good use of everything instead of buying an overload. Moreover, you will be able to use things before they are expired or do not remain fresh. This will save some precious bucks on your next visit to the grocery store.

6. **Go meatless on particular weekdays:**

Meat is usually one of the more expensive items on your grocery list. Specify a few days in the week when you go meatless and replace it with some other good protein source. When we discuss meat, one thing more that you can do is replace premium meat cuts with cheap meat cuts. These can be used in a variety of ways to add flavor to your food. Enjoy!

7. **Invest time in preparing food when you are free:**

Skip fast foods, microwave meals, and so on. Instead of eating fast food or just munching on some prepackaged food when you get home, try making some simple and healthy replacements that you can eat. An hour's worth of preparation one weekend can give you a ton of cheap and handy meals that will end up saving you a lot of cash and not eat into your time when you're busy. If you have a very busy work week, try to prepare your food on weekends in advance.

8. Try to plan your trip at a peaceful time:

Although sometimes it is not practical to go to the grocery store alone, try it when you can. You will be delighted with the amount of savings. Kids usually add on a lot of unnecessary things in their shopping cart. Also, you can prove to be a culprit yourself if you feel hungry or tired. A peaceful trip to the store can result in smart shopping and great savings. Try to reduce the number of grocery shopping trips. Not only will you save on fuel, fewer trips to the grocery stores mean fewer chances of buying unwanted things. Double savings!

Save On New Tech

1. Don't jump on the upgrade bandwagon:

Suppose you scrutinize the sorts of technology acquisitions and enhancements you've created. In that case, you can see that most of them were needless. And, even though we all purchase products to fulfill an urgent requirement, we spend on new technology because of peer pressure.

More often than not, we just want something that has the latest and greatest in business terms. Don't buy Some technology on impulse. Take a second to consider before spending your money on the next technology acquisition. Do it, if you want.

2. **Always Do Your Research:**

I just cannot emphasize this point enough. Knowledge is power, and these days, it is very easy to acquire this power through the internet.

Do your research really well, and the good part is that you do not have to be an expert at it.

3. **Don't equate price to quality:**

Sometimes, we tend to get so impressed by the high price that we automatically perceive it to be of great value. Don't do that. Expensive

does not necessarily mean high quality as well. Go for a product that actually fulfills your needs in a great manner, and it does not have to leave you with a hole in the pocket.

4. **Don't go for an extended warranty or insurance:**

The seller will definitely offer you an extended warranty or insurance. This will naturally raise the cost of the product and most probably will not be worth it. Skip this offer.

You will be better off by creating your own fund for similar future purchases in the long run rather than relying on the extended warranty.

Save On Vacations

All work and no play make Jack a dull boy! And we surely don't want that to happen to you. You should take vacations and spend quality time with your family.

The only problem with this is usually the finances. Trust me; you do not have to wipe out your bank account to have a great time on vacation. Make smart decisions, and you can get a great deal for your money.

1. **Skip the travel agent:**

Although they save you a lot of headaches, they leave you with a dent in your pocket too! Try doing the dirty work yourself. That will save you a great deal of money.

2. **Use the internet to find the best deals:**

Gone are the days when travel agents used to be the only ones privy to great deals and discounts. With the fast-growing use of the internet, there are many websites through which you can search for the best of deals and discounts yourself. They are a dick away. You can try using different websites to find the deals once you have decided on your vacation destination. Amazing savings can be done from flights. Also, make use of websites that guarantee the lowest rates by comparing different services or brands.

Websites that offer price comparisons sometimes seem to be a waste of our time but can provide immense value on many services related to vacationing, such as on car hire, travel insurance, and package holidays. They can certainly save you money. Although, I will suggest that you should look closely at all the choices. At many times, some hidden charges or taxes can rip you off your valuable money.

3. **Travel when it is the cheapest:**

Flight fares vary considerably depending on the time of the year you are traveling. Even within the same month, ticket prices can be vastly different on various days or even flight times. Look out for the cheapest tickets. If you are flexible on your travel dates, you can get great savings on your tickets; one of the major vacationing components.

4. **Don't overspend on your accommodation:**

Obviously, you are not going to spend much of your time locked inside your hotel room. You might want to spend your hard-earned bucks on amazing activities instead. You can also save a lot of money if you can find alternative accommodation to a hotel room such as a home swap, an apartment, or a private room.

5. **Save money on food:**

No vacation is complete without tasty, local food! The good part is that the best food options are usually the cheaper ones. Of course, if you eat at a fancy hotel, it will be more expensive than the street food and will probably not taste that good.

Restaurants and cafes at some distance from the big sights are definitely cheaper than those whose main trade comes from tourists. A five-minute walk can halve your bill and get you a much better (read: delicious) experience.

Avoid opening the minibar. A seemingly harmless drink or snack can cost you a lot of money. Instead, if you absolutely deem munching as an indispensable part of your vacations, stock up on such items when you go out of your hotel and carry them with you.

6. **Avoid the extras:**

Okay, so your planned the trip at the right time, got the best available tickets, and everything is going in the right direction but wait, have you weighed your bags yet? Sometimes, the payment for going over the limit can eliminate your savings and put an unnecessary hole in your pocket. Be careful.

Whether you are planning to travel light by taking hand baggage only or carrying quite some luggage, be absolutely sure that your bags do not exceed the weight limits imposed by your airline for your particular ticket and check it well before you leave for the airport. Excess bagging charges are the last extra thing you would want to pay, especially on vacation.

7. **Save money on traveling when you go for a vacation:**

Instead of going for a pre-arranged excursion, plan your own. That results in cutting the frills and costs. Usually, the trip arranger charges a premium for taking the pain, but why not try things out yourself? After all, you are on vacation and have time on your hands to enjoy.

Also, look out for the cheaper transport options. It is usually cheaper and more interesting to move around by bus or any other locally available option than hiring a car and paying for all the additional costs like fuel, car insurance, etc. Do you want an even better option? Try walking! I am absolutely serious. You will discover a lot of places that you will otherwise miss. Skip the next bus ride and start walking. You will have fun and save a lot of money.

8. **Beware of hidden costs and charges:**

Whenever you are away from home, things change. You might be facing extra charges with the things you do not think about twice when you are in your home country. One such example is roaming charges on your mobile phone. You need to check your tariffs. Some contracts allow people to use an allotted home allowance for a charge of certain money a day. But this can be a very bad deal. If all one has to do is to send two or three texts a day and use Skype to call home in the evenings, they can change to another tariff and pay very low charges. Again, make use of free Wi-Fi instead to make calls, use Skype or check your email. Many hotels offer free Wi-Fi these days.

Saving On Healthcare Costs

One of the most important and unavoidable costs in our life is probably the cost related to healthcare. Whether it is for our own self or family, we do not want to make any compromises when it comes to health. You don't have to, but that does not mean that you have to go blindly by your doctor's or insurance company's words. Check out the following tips:

1. **Choose your health plan wisely**

When choosing among multiple insurance policies, don't just go for the one with the lowest monthly premium or one you're already using; you must look for a new plan that meets your specific needs. Most of the benefits transition will occur as the situation and needs change.

An all-time great weapon! You should jot down your family's total number of medical visits and the number of medications they receive each year, as well as other services, including shots and check-ups. Compare the annual costs of both of these plans to one another. You can also factor in the annual premiums and co-payments.

2. **Inquire about generic medicine:**

Always ask your doctor for a suitable generic option instead of a particular brand of drug. They all perform the same function, but many plans will charge significantly greater co-pay if you use a name-brand drug versus a generic.

You may also want to ask if you can take something over-the-counter instead of when your doctor recommends a vitamin supplement.

3. **Compare costs when your doctor suggests lab tests:**

Some laboratories charge a hefty price for the same test that can be done at a much cheaper cost. So look around, do your research and then choose one lab. Even if your doctor recommends a particular facility for getting the tests done, ask him specifically about other options that offer good quality at a cheap rate. The costs vary hugely when it comes to testing facilities.

4. **Check your bills:**

We all are prone to errors, so are the doctors, labs, and pharmacies. Incorrect or erroneous billing might cause a dent in your pocket. Always recheck your bills to see if the billing is correct. There's no harm in checking and rechecking.

5. **Check out the drug manufacturer's website:**

You can get some coupons or discounts through a discount card when you are using the drug manufacturer's website. It will automatically result in more savings, especially if you are a regular user of a medicine.

6. Maintain a healthy lifestyle:

You might find this point irrelevant in a discussion when all you want to know is how to save more. This is exactly relevant to savings. If you maintain a healthy lifestyle, walk daily, eat healthy (which by the way is cheaper too), and in general practice good habits, you are likely to avoid the doctor's office and naturally save the money that otherwise would be spent on doctor's visits, lab tests, and medicines.

As Anne Wilson Schaef pointed out, *"Good health is not something we can buy.*

However, it can be an extremely valuable savings account."

7. Prevention is better than cure

This proves to be true in every situation, especially when we are talking about saving on health-related costs. Don't avoid routine checkups just to save some

money. If you do, you might end up paying a lot of extra money. In any case, routine exams are usually inexpensive or fully covered by insurance. They can lead to the early detection of health problems.

Likewise, be vigilant about your child's vaccination. Routine vaccines will also lower your child's risk of illness and naturally lead to reduced healthcare costs for you.

Other Random Tips

1. Reduce the cost of commuting to work

If you fly to work on your own means, you're likely to be an enemy of the earth. Watch out for co-workers who live in the neighborhood. Divvying up the costs will save you money. Try to find the cheapest gasoline The money you spend at each station will differ by small amounts, but over the month, it will save you much more or lose you more. Reduce your heater and AC use to save money.

Additionally, they consume additional oxygen. Just cool your engine, then switch it off and turn it off. Some advice rolling down the windows before using air-conditioning the car to bring it down substantially, or vice versa (if you must). You should, of course, make mental notes of things you have to do when on the way to the office.

2. Make your payments promptly

In general, make payments for all your bills promptly. If you forget things habitually, try to make use of the auto-payment options offered by your bank

3. Shop online

Almost everything can be found online for a lesser price. This goes for both new and similar products. It also doesn't matter if you're shopping for groceries, health and beauty products, or even a game console. Just invest some time in researching, and you can reward yourself with more savings in hand. Even better is that many companies now offer free shipping to improve their sales and don't charge sales tax.

4. Buy used items

It is not necessary to always buy brand-new. You can shop for slightly used items in good condition. There are lots of people out there who sell their goods in great condition. Make full use of garden sales but only buy what you really need. Don't get tempted by seemingly low prices. If you end up buying more than what you require, you will, in fact, lose your savings rather than increasing them.

5. Improve your credit score

Improving your credit score is certainly the most important way of saving money. A good credit score can save you a small fortune from home loans and car loans to credit cards and auto insurance. The savings might look small, but in total, you can save a lot from maintaining a good credit score.

6. Combine your bills

One of the biggest monthly expenses for some these days is Internet service, cable, and phone costs. The major providers today offer discounts when you bundle all three of these services together. This will save you the hassle of paying separate bills and earn you good savings as well.

7. Sell everything you don't need

Really! Get together with your family and clean up everything that you don't need in the house. Hold a garden sale, have fun and earn money! That will give you lots of benefits. You will not only earn money that will add to your savings but also have a cleaner house!

8. Look for a cashback card or scheme

Many websites, credit cards, and companies give you cashback facilities on your purchases. You can browse High Street and online retailers using the websites. Once you buy an item, you will receive a percentage of the value of your shopping as cashback which will be paid into your account.

9. Cancel Unwanted Memberships

This is a perfect illustration of a need versus desire. It's a gym subscription you hardly use, isn't it? That is not a requirement. There are myriad ways to keep in shape without expending any effort. Using the out-of-doors.

Explore and enjoy nearby bike trails. The 3-day weekend hike goes out to play in the park; go to YouTube and search for 3o minute videos of total body workouts; you will be surprised at the results!

Even if you hate that one title, we've found the best deals. Signing up with a friend or member of your family will save you money. It's usually better to test out a gym for a full week and see what you like before you sign up for a longer period. Ensure that you find out whether they'll waive the registration fee. Businesses have to register with the court to practice in these cases to keep their licenses, so you might as well do it.

I looked at my gym's Facebook page to see if there was a previous promotion going, and I signed up." After a while, I kept waiting and was rewarded with new member specials. I'm trying to get across if you really have to be fit and take innovative approaches to do so.

You may be not a gym rat but have other memberships or programs chipping away at your financial and physical well-being. Since these days, membership subscriptions are more likely to take the form of digital content providers or regular delivery. Do you really have to subscribe to a streaming service to view every show in the queue? Is there something in the artisanal jerky of the month club that you need to have at the moment?

10. Cut Down On Cell-Phone Bill

You're likely paying more than you should for your phone. Tackling is not the only way to handle this problem.

Organize a family vacation. When you're on an account, you may have several lines, but be sure to sign up as an unlimited data plan; otherwise, you'll be covered in fur. The cost per user is lower as more people sign up for the program. He's married to a large family, and eight people on his phone bill are cheaper than anyone in it. You're a real charmer!

Let's not add the new model until we know whether we need it or not. A phone is intended to serve for more than two years!

Enthused by the prospect of buying the new model? Paying in advance is better. Negotiating a contract is less painful in the short term and also lowers the long-term cost. In other words, when I recently bought a modem, my monthly payments came down to $30 per month. In addition, because I have no further obligation to a rival, I'm alert for ditch and turn specials.

Don't sell your second-hand cellphone. You're better off self-publishing rather than trading the rights in.

Keep your phone covered with a sturdy case. For a long car commute, the Otter Box is an excellent option. It is alright to spend up to $2 on this as long as you have chosen not to get benefits (Spoiler: I recommend doing so in Chapter 11). The chance of breaking your screen while trying to replace it yourself will void your warranty and prevent you from being able to trade the phone in for another.

Try opting for unlimited if your package penalizes data overages. You may be paying for just the bandwidth you use, but overages can be stealthy. This is particularly true if you stream videos or music while you are on the go. You will see if this is something you'll need to keep an eye on in your online resource portal.

No phone battery should be left in a hot for more than an hour in Austin, Texas, regardless of temperature. I discovered that there's no way to trade your phone in if you make a cake for it.

11. Cut Down On Internet Bill

Pro tip: Buy your own modem and router if you haven't already. I bought my modem router 2-in-1 for about $70 from Amazon while I was at Comcast signing up for my new account. Renting can cost you $120 every year!

I do not recommend buying a second-hand router or modem. There's no guarantee that it will work, and/or it may break shortly after you get it (this is one of those failure moments I mentioned that I'm skipping over in the interest of time and sounding smart!).

If you've been with your internet and/or cable provider for less than a year, odds are you are on a special introductory rate. Call them before the year ends and ask them to honor the price for the next year. Threaten to cancel your account if needed — do some research online and have a specific competitor and "deal" in mind.

Even if you've been with your provider for over a year, call them and see if they can't do anything to lower your bill. Nine times out of ten, your provider will not want to lose you as a customer and will work with you. If you have no luck, look at different plans and consider dropping your speed or reducing your channels to save money. Here are my 5 strategies for negotiating your bill:

I. **Pick up the phone:** don't waste time emailing back and forth or chatting with a bot. Call the company directly and talk to a human.

II. **Be prepared:** have your account information ready and know the price you are currently paying. Next, know the cost of the service or packages that are available.

III. **Don't accept the first offer**: because what would be the fun in that? Keep pushing for a better deal!

IV. **Be nice**: just because you want to lower your bill doesn't mean you have to be rude about it. Be kind and use the name of the person you're talking to. Kindness can go far, and many people are more willing to be helpful when they are treated well.

V. **Know the competition:** if you are threatening to cancel your service, know what the competitor offers. Mentioning this specifically might get your current provider to beat or match the price or services of the competitor.

12. Negotiate Rent

Rent. The biggie. The truth is, you should always shop around when you're nearing the end of your lease. You simply never know what else is out there. You might find a better rate. At the very least, you'll be able to make an educated decision when you renew your lease. And when you do renew, *always negotiate.* The worst they can say is "no." Don't forget, finding good new tenants can be a hassle that many would rather avoid.

Look around your current place. Do you need all that square footage? Do you really need that breakfast nook or second bedroom that you never use? Do you need those high ceilings and "old world charm"? Or do you just *want* it? If you live alone, consider roommates. If you pay for parking or storage in your building, try going without next month and see if you actually miss it. Have the option to lock in a better rate if you commit to a two-year lease? Do it!

Pro tip: Start your lease between November and February. I've been on this cycle for over 5 years. You'll have less competition, which means that as homeowners and building managers get desperate, they will often give you a better rate!

You have more control over the process, and landlords would rather work with you than have to replace you in those bleak winter months. The most expensive time to look? June, July, and especially August.

13. Compare Prices Before Buying

Grocery shopping is something you'll have to do even if you've never been able to do it before. Trader Joe's became my new favorite place for fruit, bread, veggie meals, beer, and treats, as well as my favorite grocery store for beer, food, both of which are supremely important for us! But, for some meats, yogurts, fruits, bulk spices, and basics, I will always continue to go to Whole Foods. Keep track of the pricing so you can see when a deal is offered. But when they're on sale — buy as many as you can.

Bulk bins are excellent places to save on various grain products, such as whole-wheat pasta, rice, nuts, granola, and spices. Saves money, doesn't use a lot of room, and provides a better yield, too. There are no problems with re-enablement. When you return. It should only ensure you have enough storage containers to put it in when you return it.

Want to get a cash bonus when you shop using Ibotta? Try using the reward's app Ibotta. Every time you use it, it adds a new item to your basket. Today I saved $2.50 on things that I would have purchased regardless of the price. So far, the free money has added up to a total of $166.12 in my lifetime!

14. Bring Your Own Lunch To Work

While this one might veer towards the obvious advice I promised to avoid, this part is the kicker: every day. Commit to it! Pack up your dinner leftovers. Stock up on inexpensive frozen microwaveable meals (for example, a filling butter chicken).

15. Be Thrifty

There are some pretty trendy thrift shops these days (thanks, Macklemore!). Do a quick web search in your neighborhood and check some out! Locally, I frequent Buffalo Exchange and Plato's Closet. Don't overlook Goodwill, too, for someone-of-a-kind finds from a place that also does great things for your community. Are you and a friend both tired of your own closets? Try a clothes swap! For special occasions, pay a little to get that old suit or dress fixed, trimmed, or slim. New suits cost $400, $50o minimum — have it tailored for $50 instead. Repair those old heels instead of buying new ones. Just don't break them again the next time you do the Electric Slide! Savings — "it's electric!"

Pro tip: If you find that your clothes are just too loved (i.e., threadbare, worn out beyond reasonable repair), find a textile recycling center in your area. Programs are popping up all the time that allow you to either drop off or mail in your rags to a program that then recycles the threads into something else that is useful, like new clothing or home insulation. Money isn't the only green around here! And those are just the ones on the top of my head; there are thousands of more ways by which you can save money and utilize them towards reaching your long/short-term financial goals. The key is to be Frugal in a Smart way so that you don't have to let go of all the things you enjoy in life, bit not indulge yourself in so much luxury that you end up having nothing end of the day.

CHAPTER FOUR

Effective Ways of Earning Money

This is a huge topic to deal with, especially because there are so many legitimate ways through which you can start earning money right from home and without any investment. Sure, you always have the option to develop your skills and apply for a day-to-day job, but that is really something that most people these days tend to avoid. Especially during this COVID season, more and more people are shifting their businesses online. They are developing/pursuing very interesting ways to earning money by working right from the comfort of home.

This chapter will focus on sharing some valuable ideas for you to choose and start earning money. I tried to select mostly generalized strategies that anyone who knows how to read and write will pursue!

So, whether you are a student or a professional employee who is tired of their 9-5 job, there is something for everyone!

So, Whether you've been searching for a way to increase your income when working a full-time job or taking on other projects, this eBook will get you started in the right direction.

Side hustles are essentially lucrative prospects, that you can take advantage of without devoting a significant amount of time.

While, many people begin making money with side hustles as a means to supplement their monthly pay. It is not unusual for side hustles to soon become a main source of income and overtake the 9-5.So, to start, look at the various ways to make money described in this eBook. Make notes and suggestions as you read through the content, and then make a plan of action when you're done!

Side hustles are one of the most effective ways to pay off loans fast, prepare for retirement and even reduce the financial burden, that let you can enjoy your best life. Finally, the strongest side hustles are adaptable and can prove lucrative without requiring a significant time commitment.

Although, most of the side hustles featured in this eBook can easily be extended into a much larger enterprise. They are built to allow you to work as little or as much as you want when you are first starting out. So, without further ado, let's get started!

Learning To Produce Video Tutorials Online

An abundance of methods for making money on the internet exists. You'll find many ways to succeed if you just take it seriously and think of ways to make a living. I found a fun and exciting way to make a little extra money online. Everybody certainly knows about your subject; it's just a matter of where you look for the details.

People look for how-to videos to discover everything from tasks to learn. Creating an amazing video tutorial website like Vimeo makes it easy to start a video blog and pulls in premium traffic (and at little to no cost).

First, think about the type of content you want to produce: That is, focus on providing valuable and specific content that can be converted into tutorials.

Next, you can see if you can find some relevant instructional videos on YouTube and get to know the style, voice, and style of those who already make the videos in that niche. Think about the type of final video you want to make: presentation, explainer, or screencast.

The more time you spend analyzing your audience, the easier it will be to identify new creative approaches to their problems.

It's possible to make a storyboard or script that will show you where the movie is headed while you work. This will cut down on errors and make for better TV viewership, enabling viewers to keep their attention on the program. Next, you can acquire high-quality machinery, but you don't have to spend all your money on it.

Here are our Best Recommendation

- Webcam (Best camcorders for shooting video lectures).
- Snagit (screen-capture software program) — For recording your computer screen.
- Camtasia (high-quality microphone) — For recording audio.

You only need a high-quality microphone and a screen-capture software application, such as Snagit or Camtasia, to get started. These are two of the most used techniques used by both beginners and professionals.

If you want to learn more about this side hustle, here's a perfect step-by-step guide to getting started: https://amzn.to/3tQlhfC

Provide Tutorial Or Training

You can take this a step further by developing online workshops and training programs, which are similar to creating video guides based on your experience and expertise.

Students can pay to gain access to the instruction, which can be offered in various ways, including through common course sites such as Skill share or Udemy.

When creating training material, you can create video-based videos, text-based courses, or a combination of the two.

The preferred style should be determined by your market and what they are looking for. So, spend some time exploring your niche, including current courses on your chosen subject.

Courses are a great deal! People make millions of dollars every year by providing basic classes that help people develop new talent.

Your course also does not have to be overly long. In reality, less is always enough, so students choose to spend their time studying the most important steps or components to accomplish their goals. And if you think you're not able to build a course, reconsider!

People are buying courses on almost every topic imaginable. So, start by evaluating your personal skills and desires. Then, brainstorming on few ideas that you can later qualify should be possible. Maybe you're excellent at bookkeeping, or you know how to design websites or make videos.

All of these are useful qualities that others will be willing to spend decent money to master.

Tip: This is one of the most lucrative side hustles in this book, but you can make much more money or extra fee with the additional offer of counseling, selection of student's ability to get personal support and coaching.

This is a tried-and-true form of increasing your earnings and helps you tailor the upgrade in various ways, including by providing different levels. For example, you might have coaching for 30 days, 60 days, or encourage students to buy time blocks. Alternatively, you could have a subscription plan in your course that allows students to pay for access to additional tools and services.

Popular industry and its marketing by creating content pages on a blog for the informative feedback will assist buyers in their buying decision.

Become a Freelancer

Freelancing is an excellent way to supplement your income without devoting a significant amount of time to the efforts. Simply create a service menu, set your prices, and take on just as many jobs as you have time for each week.

And since freelancing comes in several ways, based on your abilities, you'll be able to easily create a client base by simply building a profile on platforms like Upwork, freelancer and fiverr.

Writing, programming, architecture, virtual assistants and specialist fields such as legal, translation, and mobile creation are all available.

Tips: Be certain that you only take jobs that you are certain you will fulfill. User input is critical on freelance platforms like Upwork because it lets potential clients assess your dependability.

This ensures that you constantly strive to reach deadlines. If you anticipate any problems, you email your client as soon as possible. After the project is completed, regular contact is critical for promoting constructive reviews.

Furthermore, do a market analysis to set fair pricing points on the different services you have.

You never want to undervalue your job or undercut yourself, so research what other freelancers in your industry are charging for similar services.

Here are a couple more freelance marketplaces to lock into:

- Freelancer: https://www.freelancer.com/
- Guru: https://www.guru.com/
- Toptal : https://wmv.toptan
- We Work Remotely: https://weworkremotely.com/
- Dribble: https://dribbble.com/jobs
- Fiverr : https://www.fiverr.com

Looking Into Dropshipping

Dropshipping allows you to distribute a product while leaving you without providing an inventory or purchase inventory.

You can submit it directly to your client, which frees you uptime and eliminates your workload on the process.

Real automation is provided by Dropshipping. It is a low-risk approach since the inventory items are not bought until they are required, and products are only shipped when requested, excluding the high-product cost (and low-involvement) inventory acquisition is possible.

Let's start by agreeing on a market, choosing a supplier, and going out to different consumer demographics using distribution channels such as Amazon or eBay.

Your focus is essential to hone your research and determine what kind of product or service you want to sell.

Our aim is to concentrate your focus on year-round income-generating products.

Focus on hot trends, but don't dedicate your business to them. Instead, choose busy industries that have a constant flow of repeat clients.

Resource: This website is helpful if you want to stay up to date with the latest trends: https://www.trendhunter.com/

And if you want to get started in traditional, evergreen niches, the Oberlo app will be very useful: https://www.oberlo.com/

Finally, you can locate vendors online, such as Oberlo, and easily create a shop that incorporates their inventory.

We recommend:https://www.shopifv.com or https:-ww.bigcommerce.com

Sell T-Shirt And Other Merchandises

Did you know you can make money selling t-shirts without even having to ship merchandise or even process a payment?

With websites like https://www.RedBubble.com and https://printifv.com You can build and submit templates, and they can do the rest.

You will be paid a sum on each item sold and will be able to choose from a range of product categories to display the graphic on.

This ensures you can sell items other than t-shirts, such as coffee mugs, mousepads, and even handbags.

This is a perfect side hustle because it doesn't need much time or commitment to get started, and you can simply work on new projects in your spare time.

You should spend some time studying various markets to develop unique concepts that are sure to sell.

This is often the most time-consuming move, but it is well worth it. Classic brands can be found on websites such as Amazon and Etsy. Take notes on common words, lingo, and even color schemes that are selling, since this detail can be useful when designing your own projects.

Of course, suppose you're not an artist. In that case, you can always get in on the action by recruiting graphic artists on sites like Fiverr.com. Still, luckily, sites like Canva.com make it very easy to produce graphics for t-shirts, even if you have no experience!

Tip: Make sure the graphics are 300 dpi and saved in.PNG files of 4500x5400 dimensions.

Look Into Self-Publishing

The Amazon KDP (Kindle Direct Publishing) Platform makes it simple for everyone to get published without submitting question letters to publishers in the hopes of being picked up.

Indeed, with a free Amazon KDP account, you can publish a book in various markets, including fiction and nonfiction.

Link: https://kdp.amazon.com

To begin, consider the various topics in which you are personally involved and have experience.

Are you a fan of historical, romantic, or fantasy nonfiction? Consider writing a short story and then expanding it into a full-length book or series after you've had your feet wet in publishing.

Are you familiar with traditional nonfiction subjects like self-help, industry, finance, or personal development? What are some of Amazon's best-selling categories?

Even if you aren't a natural writer, don't let that deter you from doing this lucrative side hustle.

You can quickly build a plot plan and outsource the writing to ghostwriters on marketplaces like https://www.UpWork.com , https://www.Reedsy.com. and https://shop.nicheraiders.com.

Be sure to describe your project priorities when making an outline for a ghostwriter. Make your vision as precise and straightforward as possible so that your ghostwriter truly understands it.

Then, before making contracts or submitting payments or milestones, evaluate the ghostwriter's abilities. It is important to employ a ghostwriter who has the voice and style you believe are perfect for your plot. Reviewing a ghostwriter's biography, especially previous works in your genre will help you determine if they are a good fit.

Self-publishing is very lucrative. Suppose you set up a ghostwriting and editing team. In that case, you will simplify a significant portion of the process, allowing you to easily build a passive income source.

Hot Tips: Don't end with Amazon. You will also make money by writing short posts on blogs like Medium. Corn and short "episodes" on famous websites like http://radishfiction.com, where some young writers are making six figures a year for short stories! Even bearing in mind that self-publishing isn't exclusive to fiction and nonfiction novels!

You will make money by publishing "low content" books on Amazon. This includes items such as planners, schedules, pens, and others.

These not only sell all year, but they also take very little time to make once you get the hang of it.

You can also save time by ordering pre-made planner models and low-content books with absolute commercial-use privileges from websites such as shop.nicheraiders.com

Make Money For Your Talents

Surveying your perspectives and skills is one of the easiest opportunities to discover new side hustles.

Many popular companies are founded on someone's personal hobby or expertise — which isn't always apparent. This is why we recommend making a list of anything you are excited about, are interested in, or have expertise or experience in.

The options are infinite if it's gaming — which could be transformed into a lucrative Livestream on gaming sites like Twitch — whether you're an enthusiastic reader and could provide proofreading or writing services to new authors.

Once you've compiled a list of potential prospects, you'll need to qualify them by spending some time investigating each one to develop a game plan.

See the same:

1. **Market demand:** Is there a current market for such goods and services?
2. **Availability & Pricing:** Consider the various formats and price points that are currently available online. Consider how you could put the bid to be successful.
3. **Time Required:** You should also consider how much time it would take to develop a product or service in this niche and whether you have the time to devote to it.
4. **Difficulty Level:** Determine whether the product or service you want to create is something you should do yourself or whether it can be outsourced to skilled freelancers on platforms like Upwork.

Set a start and finish date and get started after evaluating the competition and having a good outline that will give you a sense of direction.

Tip: Look for and visit Facebook groups or Reddit posts about the niche you're involved in. It's a smart way to study the niche and get in line for potential networking opportunities.

Exploring Amazon FBA

People generally regard Amazon as the world's most successful eCommerce company (Amazon.com). FBA is a combination of both wholesaling and retailing for them.

When selling on Amazon using FBA, you get the advantage of their free stockpile, bunding, shipping and logistics services. Unlike outsourcing, where you own no product, have you invested time or money in Amazon Fulfillisteria? All is packed and ready to go, all thanks to the generosity of Amazon.

It works to your benefit because customers believe they are getting good stuff from your site due to your association with Amazon. Because of this, you'll never have to care for or move massive amounts of gear. Assuming you use Amazon FBA, you would likely pay lower shipping expenses.

As I'm very certain you have noted, using Amazon's administrations would necessitate a little cash investment. You have to pay a 15% of the item plus shipping and handling for Amazon to collect the item whenever it sells.

Outsourcing now seems to be attractive because it allows you to pay less upfront. Still, each has advantages according to the strategy.

You must build a customer base to commercialize your product. But at the same time, Amazon would not prevent you from extending your revenue base because of their generous client base.

Normally, you will have questions involving major success stories, like Amazon, but this is a good idea for smaller brands, too. The expansion of the procuring capacity of Amazon's total sales and income sources is not outside the realm of possibility.

Spencer Haws, the owner of Niche Pursuits, listed his products on Amazon for around 40% more than their potential net worth to realize a profit of 10,000 dollars in his first month of operation. That doesn't you see; there are enormous gains to be found in tracking down a specialty. The prospects that hold for eternity!

1. Building Your Amazon FBA Business

So since you're keen on beginning your own Amazon FBA business, where do you really start?

- **Open an account and consider your niche**

Indeed, the initial step is to open a 'Vender Account' with Amazon. You can either select a Professional or an Individual record. The Individual record doesn't need any month-to-month membership; anyway, you will have a few limits set for you. These could work for you or against you, contingent upon the requirements of your business.

Similarly, as you would do in outsourcing, you should discover a specialty for your business. Make sure to relate it to something you're keen on, like a leisure activity or unique interest. Try not to discard additional thoughts you may have, as you might actually add them later on, or use them as another option if your present specialty isn't ending up being effective. As usual, grow your insight however much as could reasonably be expected, with the goal that you can more readily decide the most ideal results for your business.

- **What products will you sell?**

From that point, you should figure out what items you might want to sell. Consider utilizing catchphrase apparatuses to help you accumulate the investigation into how regularly individuals look for specific items and utilize this to figure out what are the best items to sell.

Make a point to recognize what charges are related to every item. Weight, size, and capacity prerequisites will differ contingent upon your item, so search for items with lower charges to limit your costs.

Give a valiant effort to avoid items utilized by mainstream brand names and are grounded in the business sectors. These are boundlessly harder to rival.

- **Product Sourcing**

Whenever you've decided the center items that your specialty will base on, you should discover where your items will be sourced. You ought to be searching for provisions with excellent items accessible for you to sell.

You could go to your neighborhood producer on the off chance that you expect to sell an item that is created locally. I'm positive they will be inviting to your efforts. Alternatively, you could take the course of utilizing an abroad provider, which is likewise useful as it will, in general, bring the potential for higher overall revenues for your business. It would be your own advantage to test the items you are anticipating selling, should you wind up sourcing from an abroad provider. Attempt to get your hands on certain examples so you can guarantee you will give top-notch things to your future customers.

Your following stage will be to deliver the products to an Amazon Fulfillment Center. Amazon will furnish you with details and directions while going through this interaction. Make certain to remember dispatching costs, as these can wreck your business if you neglect to monitor them.

- **Creating your brand**

You presently have items fit to be sold, be that as it may, you're actually going to require a brand. Your image will tell your customers what your identity is. Make a point to investigate if your image name has effectively been utilized or not, just as whether there is an accessible space. Your area name is your site's name, and this will be the place where customers can discover you.

Even though you acquire a brand that will acquire believability through Amazon, it is significant for you to put a high worth on its creation. This will characterize your business however long it exists.

Thus, you need to ensure it gets going on the correct foot. Recollect the fundamental aspects of a brand are its name, slogan, logo, and brand story. This will likewise incorporate your items and how you decide to show them.

Make your item pages, ensuring your item subtleties are exact and definite. In contrast, item photographs are proficient that accentuate and flaunt your items. Investigate how different brands market their own items, utilize these ways to control your own promoting.

Ensure the tone of your portrayals are amicable and alluring to expected purchasers. 'Sell' the item, yet don't allow them to feel that you are selling it. Utilize the portrayals to cause customers to feel like this is something they should purchase that they can't survive without. Be expressive in your composition. Ensure any sections you have are organized a lot, so they are alluring to the peruser. A lot of boxed altogether will feel somewhat overpowering.

The point of the portrayal is to sell the item and to quiet all questions a customer may have concerning their buy.

Marketing

The FBA settings are complete. Now, we must do so. It is important to get your company ready for action and your deals into position, regardless of whether you decide to outsource your work to other people or keep it in-house.

Still, for now, it is your best friend on the Internet. Have a look at raising your profile, allowing people to see, tag, and share your images. Preparation for possible clients.

As you can, focus on highlighting a few channels without pause and try to allocate an equal amount of time to each. There is no advantage to picking two or three channels at the outset; once you've achieved a perceived success, you can move on to other channels.

Influencers are useful because they can help you market your product. GIVE AWAY your good products in exchange for exhibiting and marketing your merchandise, and you can see your business grow quickly.

Although, all this is advantageous, it's also critical to have stages on the platform and web-based media for consumers to leave feedback on their purchases and administration. They'll gain significant favor for you with their current customers. Still, you will also get an amazing benefit from your clients. To maximize the number of people who want to subscribe to your email list, cultivate your existing and prospective clients.

Using your platform to interest consumers but never appear to push them over the edge and force them to join should be a goal of your membership strategy. Jokingly

2. **A Few Extra Tips**

Make it a steady objective to improve your Best Sellers Rank (BSR). The term being genuinely plain as day, having a business with predictable BSR development will show customers that your business can succeed and dependable on the quality you offer.

Stay in front of the patterns, and give more proposals to your customers at whatever point conceivable. Keep on including items that will praise the ones you have yet give more assortment in choice.

You could likewise keep on expanding your incomes by turning into a subsidiary with Amazon. Nonetheless, we'll examine subsidiary promotion in detail somewhat later.

With Amazon's broad image and instruments accessible to you, there is a universe of unlimited prospects that anticipate you pick Amazon FBA to additional your endeavor.

Exploring The Concepts Of Affiliate Marketing

Presently, as recently referenced, we'll move onto associate promoting. If you've known about associate advertising previously, I envision you're figuring it could be confounded and not worth the exertion. In any case, on the off chance that you've done any earlier research on easy revenue already, you'd realize that associate promoting is viewed as one of the go-to promising circumstances with regards to easy revenue.

1. What is affiliate marketing?

Associate showcasing is the interaction with which you can acquire commission by advancing anything created by someone other than ourselves. Even though income is acquired through various compensation plots, the goal is to make any purchase result profitable. It takes the sting out of the phrase "get paid while you snooze" by implementing your subsidiary code and bringing its soulfulness. You could make a couple of dollars, or you might make millions. It's about getting ahead and using your advantages to sell your product.

In contrast to outsourcing and fulfill Your Basket at Amazon, you are not directly concerned with the goods, procurement, shipping, or receiving. You're the only one responsible for the increased sales of that item. This isn't to say you should be an associate for anything, but it does mean you should turn down several assignments. You do not simply deal with an object, as you both deal with the customers. The sound of confidence here is to make it even more certain that you can win the loyalty of your customers.

No matter if you've never been to this site before, you've certainly heard of TripAdvisor before. TripAdvisor is an online travel platform meant to accommodate discounts, service charges, and restaurant recommendations for lodgings worldwide. Never get predictable; instead of getting better, learn to expect the unexpected. TripAdvisor is a member of our Creative Circle! You couldn't have even used it, but it regularly gets 1.6 million visitors. They may be one of the most successful examples of a creative and business partnership in the advertising world.

Not every person who believes he or she has achieved a great goal has guaranteed success. Regardless, you will get paid extra; it will merely augment your reward and ensure that you can live in prosperity.

Remember that we're willing to participate in something that does not fall under the banner of easy money schemes. We mean to create a state of things in which we can go on with our lives while enjoying our dreams. By the end of the passage, the reward is a great risk.

2. Creating an Affiliate Marketing Scheme

Likewise, with the entirety of the types of automated revenue we have taken a gander at up until now, you should discover a specialty for which you can turn into a member. You may be asking, "If it's simply promoting, for what reason do I need to discover a specialty?" Well, for one, it gives your site more association and structure. It permits you to draw in a market instead of going for the overall population.

When individuals are looking for a specific item or administration, you will need them to be pulled into your site, and from that point, to go on and buy something. This is the upside of a specialty market. Recollect the case of TripAdvisor? They give different alternatives and ideas to hopeful voyagers to consider when arranging their vacation. That is their specialty.

For this situation, it is shrewd to likewise consider whether there is space for another subsidiary advertiser inside your picked specialty. Very much like you would, in general, try not to enter an item market that contends with the biggest brands; it is desirable over try not to enter a swelled member market. Is it safe to say that you are ready to give rivalry and make a return? Assuming the appropriate response is indeed, definitely, take the plunge! Consider a specialty that has a wide space for associates, so your alternatives for income stay as sweeping as could be expected.

3. Research is key

From that point, the time has come to begin investigating associate projects to translate the items you will use to advance. Try not to fear taking as much time as necessary in settling on the correct decision, as it will be definitely worth your while eventually!

Search for programs that have better yields and are more productive. Remember that actual items ordinarily create lower commission rates than educational items and administrations. Notwithstanding, every last one of these can definitely be similarly as beneficial as the other. You need to consider what is appropriate for you.

For instance, if you wind up advancing actual items, joining Amazon's member program gives you a 24-hour commission reward. This implies that any item sold in the 24 hours after a client clicks your connection will give you the commission. Others have long projects, from a month to 90 days. It's dependent upon you to decide the correct fit.

4. Create your website

Presently it's the ideal opportunity for you to assemble a site. To start, you should buy an area. Space is the location of your site. Namecheap and Dream Host are two alternatives to consider; anyway, there are a few more accessible. Search for an alternative to buy that meets your monetary limitations.

Whenever you have done that, it is presently an ideal opportunity to buy and arrange a host. The host is where you will discover your site. It is the place where the total of your site dwells. Both facilitating and space buys are genuinely moderate, so don't be reluctant to push ahead, as neither one of the wills be too exorbitant if you track down the correct choice for you. Even though you should think about your monetary capacity, try not to go for the least expensive alternative for the web has. Search for solid, quality facilitating stages like BlueHost or HostGator.

Whenever you've set up a host, you will require a Content Management System. CMS, for short. CMS is programming given to you, permitting you to make and deal with a site with no requirement for specialized, inside and out information on web advancement or plan. This smooth out the way toward making a site. You don't need to become familiar with such programming to begin fabricating your site.

If you're a novice to building a site utilizing CMS, consider utilizing WordPress as it is easy to use and genuinely direct. Nonetheless, the decision is yours if you'd incline toward another stage. Attempt to keep it straightforward in your creation, as you may get yourself overcomplicating things if you're attempting to add an excess of visual allure. A basic topic will do.

5. Website Content

The following stage is to make content for your site. Your substance should stay applicable to your specialty while staying fascinating and drawing in with your intended interest group. Your substance should be sufficiently intriguing to hold your current crowd while still getting new individuals. Item audits can be valuable to furnish your customers with help with figuring out the best items for them to consider. You could be giving examination cites as a method of perceiving how one item coordinates against the other. Consider times when you were uncertain what cell phone to buy. You most likely looked online to contrast the gadgets with see which would be awesome for you.

Another option is blog posting. Sites guarantee that you have a reliable construction to building content on your site. Alongside the entirety of this, consider doling out free educational items, for example, an email membership arrangement or a digital book. This can assist with creating more interest in any item you are attempting to sell.

6. Building a follower base

When this is finished, it is currently an ideal opportunity for you to construct your crowd. On the off chance that your substance is of the greatest level, you will at last beginning bringing traffic through to your site reliably.

Here, follow the very strategies that you would consistently utilize.

- Go to online media. Another potential alternative is to utilize another person's crowd to assemble your own.
- Consider composing content for a couple of websites whose traffic is genuinely high.
- Make sure to ensure that these online journals are likewise pertinent to your specialty. You will accumulate less fascination if you choose to draw in with a unique market, which isn't solid for your possibilities.

If you have the monetary ability, you might need to think about putting resources into paid promotion. Promoting in the correct spots can improve the traffic to your site. You may likewise think about the possibility of SEO. Search Engine Optimization (SEO) is the way toward endeavoring to help your internet searcher's positioning. What's the significance here? Indeed, suppose you end up searching for a site to buy computer games on the web. In that case, you will normally utilize an internet searcher. When you do, I am positive that you don't filter through the many pages that are doled out to you. Rather you likely glance at the connections found on the principal page. The point of SEO is to be on that first page.

Consider that briefly. Envision that your site is quick to appear when individuals are looking for something identified with your specialty. This might actually drive your traffic through the rooftop!

Consider recruiting an SEO expert or setting aside the effort to learn it yourself, utilizing their strategies and standards to construct your crowd. (Stare, 2019)

7. Promoting Your Offers

Since you have developed a crowd of people, the time has come to launch your advancements! You have effectively shown individuals that the substance you give is fascinating and pertinent to them. Presently they will hear you out. The advancing substance should be possible in an assortment of ways, contingent upon you and your site.

One of these methods is "In-text joins." While your crowd is perusing your substance, they may run over connections to specific items or administrations. Should they choose to tap on the connection and buy that item or administration, you will commission on the deal. In-text joins are an unpretentious yet viable method of advancing offers, as they are organized around your substance and guarantee that your crowd isn't being besieged with an attempt to sell something, which they may not appreciate.

Another alternative, referenced beforehand, is item surveys. When your crowd will tune in to and trust you, they will be seriously tolerating and dependent on your assessment. The thought here isn't simply to call attention to the entirety of the positives and disregard the negatives. You should speak the truth about your own insight of the item and your musings of it by and large. It would be a better value if you are explicit and portray your offer and item in incredible detail. From that the point, you can give your partner a connection to the offer, and should it be bought, you've made a deal!

Member Marketing is generally about quality and procedure. The methodology you have to create quality substance and market your partner items from that point. You'll see that the absolute last thing referenced was making a deal. That is because we began with the arranging and improvement. From that point, we hoped to build up the quality expected to draw in individuals. At that point, it was tied in with advancing our subsidiary items successfully.

When that occurred, it would now be feasible to make a deal. You will track down that most of these means will be rehashed reliably all through your endeavor as a member advertiser. In any case, the deal ought to be the farthest thing from your psyche. Allow it to happen normally, as opposed to attempting to constrain a deal.

On the off chance that you follow the means and explore the business sectors inside and out. I have almost certainly that extraordinary achievement can be accomplished on your street to genuine success.

Earn Through Blogging

Not at all like a significant number of other easy revenue sources recorded in this book, is publishing content to a blog, one of those that many people have been very much aware of, however never fully found out excessively a lot.

1. What is a Blog?

Basically, a blog is such an online journal. These online journals can be utilized for the individual or drawing in with others as a more open use. These online journals can incorporate pictures, recordings, or outright antiquated content.

Begun in 1994, websites have gone from one solidarity to another in a wide range of classes, from systems administration to aggregate diversion interests to publicizing and advancement. In contrast to numerous different parts of the Internet, writing for a blog has enhanced its employments rather than being upstaged by different stages. Out of around 2 billion sites on the Internet, a huge number of them are websites. One writing for a blog stage can have around 400 million sites and clients alone, showing their unbelievable development and prominence.

As a blogger, you will have the opportunity to work at whatever point and any place you need while bringing in cash from it. Basically, publishing content to a blog is enthusiasm, and cash can be an additional advantage.

Writing for a blog enables us to do what we love while bringing in cash for it. That is what we need, correct? Obviously!

In any case, how could this be utilized to procure any pay? All things considered, that is totally reliant upon you. As referenced before, including partner offers inside your online journals can assist you with procuring pay while playing out a pastime and sharing your advantage.

An option in contrast to this is to sell publicizing on your site. This will, as a rule, expect you to have an exceptionally huge stream of traffic to have such a significant return. Generally, I would propose that this go about as an enhancement to another wellspring of easy revenue.

By and large, with contributing to a blog, the thought will be to join with different types of easy revenue to make the income. The first model would be outsourcing. You could offer items applicable to your blog for others to consider for procurement. So if you have a blog based on cosmetics, you could offer beautifying agents for others to use as a method of producing pay.

We'll talk about two different other options in more prominent detail later in this book, are selling digital books or online courses. This may imply that your underlying association might be significantly expanded. However, the awards for it are progressively defended, particularly since there are zero costs engaged with the making of these items.

The thoughts you use will rely upon the kind of blog you need to make and your abilities that will profit your business.

2. **Your Own Blog Site**

This means you've decided to start your own blog. This may be shortsighted, but all things considered, I'll come out sounding like a broken record, once again. However, on this occasion, things are a little unusual. Previously, you had made a decision based on your own preferences, which benefited you and your views; this time, it's based on your abilities and views.

Today there are more than 8 billion human beings. Anything you do is a challenge, no matter how you approach it because it still has to be done creatively. If the universe is still level, to you, in the unlikely event that it might be, you will find several people who are on the same page as you for this grand adventure. In other words, you might prove this if you find any evidence to support your claim that aliens are on Mars!

Since most of us will be unremarkable, you have your pick of a specialty where there is considerable interest. It's about finding out how you're going to get your audience excited. However, we will be there by the time you reach that conclusion. The first step to getting the ball rolling is to put up a blog.

This should be your first question. From a content perspective, obviously, there is no sense in even thinking about picking the necessity of acquiring blog-publishing capabilities unless you also pick the right to take advantage of the content. Think about what kind of blog you want to create, to begin with.

Is it going to be heavy on text or light on imagery? For example, if you wanted blog posts to be presented in text form, you would choose Tumblr. Think about Medium, assuming you wanted anything more of a blend.

What kind of involvement have you had in maintaining a blog? Will you cope with a fully customized experience, or do you prefer a simple one?

Is there any way writing for a blog can't be changed? This is mandatory, so you don't have an accurate idea of how your revenue would compare to your goals.

What do you want in payment? Some writing for a blog can cost money, and others are totally free regardless of your status as a member.

Finally, before you commit to choosing, ask yourself the following questions to avoid having any unpleasant surprises.

Like making your own website, you would need a domain name and web hosting that enables the web page to operate. The consistency of your life has already been found; it's just a matter of finding out what you want and what works for you and using that.

3. Building Your Blog

The subsequent stage is to design the subtleties of your blog.

What is the point of your blog, and how long would you say you will give to writing for a blog? The point is something you presumably definitely know, given you have decided your specialty too; however, what's more, you need to consider the measure of time that you have. In truth, you may make some full-memories work, kids, or numerous different variables that keep you involved. Contributing to a blog has to work around this, guarantee that you are as yet coordinated in your everyday life, and will not leave this behind when you feel outstandingly overpowered.

What sentiments would you like to summon in your crowd? This assists with deciding the tone and setting of your blog. It will guarantee that you are finding and composing the correct material that revolves around the point you wish to make.

Also, ultimately, how might you figure out how to draw your crowd into utilizing any of your picked income streams? Take a gander at how you will intend to advertise these streams and what steps to take to convince your crowd into a buy. This ought to be the last interesting point, as it will be based around the substance you wish to make, thus guaranteeing quality substance should be your chief need.

Presently you have an arrangement, and the time has come to plan your blog on your picked stage. It's OK to be enticed to go crazy on the creation; anyway, you'd do well to keep it easy to begin. As your experience develops and you have thought of utilizing plan as an engaged mark of fascination, you can change any way you see fit.

For the present, it simply should be flawless and pleasing. That, in itself, is intriguing to an expected crowd. A jumbled blog will simply drive them away. Take a gander at it from the crowd's point of view that you plan to draw in and move from that point.

Whenever this is done, it's an ideal opportunity to get composing! This part is totally through your own effort.

4. **Promotion is integral**

revenue together, even though I wouldn't recommend doing that, as you don't need your blog to appear to be a lucrative plan. Bar the instance of promotion space, your pay structure should enhance and be pertinent to your substance and the other way around. There is an immediate relationship between the two.

With the correct mentality, publishing content to a blog could be quite possibly the most captivating and pleasurable endeavor that you might actually inundate yourself in.

Since you have a blog fully operational with a couple of posts, it's an ideal opportunity to advance your blog. Now, I'm certain you know the rudiments of advancing a couple of types of automated revenue. On the off chance that your blog contains pictures, think about utilizing Pinterest and Instagram to advance pieces of your blog, so others will be quick to see more. As we referenced with member promoting, think about utilizing other more famous online journals to develop your crowd. Use email promoting as an approach to advise your present peruses of impending posts, to guarantee they keep returning for additional.

5. **Monetization**

Presently you need to adapt your blog. As I said before, there are a few different ways you can approach this. If your blog has the potential for pertinent items to your specialty, consider outsourcing or subsidiary advertising.

As your blog keeps on filling in ubiquity, you may start to get offers from publicists, searching for their own chances. You could decide to sell them promoting space straightforwardly. An option could be to utilize Google AdSense, smoothing out the interaction and eliminating the immediate substance. We'll get familiar with Google AdSense in a forthcoming section.

The advantage of a blog is that you can consolidate a few parts of automated

Learn Social Media Marketing

Ok. Online Media. In the current age, I don't perceive any of us living without it. Helpful from various perspectives, from staying in touch with tragically missing loved ones to creating and contacting your business crowd. As you've seen up until now, online media is assumed a major part in any undertaking you go to. It is the thing that is mostly opened to everybody without any problem.

1. So what is social media marketing?

Venturing into the web-based media promoting world is venturing into an immensely growing industry, so you should be set up to place in a fair measure of work and have a special interest as a more powerful decision than different advertisers in the field. To begin with, we need to develop your business.

2. Have a business plan

To start, you need to build up your field-tested strategy. You may have to invest some energy putting resources into yourself, as though you haven't done any sort of web-based media promoting previously, you would truly prefer not to head into this flying visually impaired. Broad exploration will be required. Consider doing a couple of free courses, just as hardly any paid ones too. When you believe you have accumulated as much information as possible about the business, the time has come to step in.

Consider building an arrangement of administrations that you might want to offer your future clients. This will give different brands lucidity on the procedures you will use in the administrations and arrangements you intend to offer. Can you offer examination or substance showcasing, or both? Attempt to discover offers that your customers would not discover at other online media organizations, to separate yourself.

3. **Professional Links**

They'll have to discover you. This should be done like this. While reading this book, I'm sure you've become aware of the pattern of site development over periods. Of course, the site needs to include the subtleties, their prices, and the contact info. Instead of focusing only on what you can do, think about why you would be a better alternative to other businesses looking for an office than theirs.

When you have finished building your website, you will need to create your own media-online accounts. You must have a strong quality for both online and offline media; if customers detect that you are static, they will stop trusting you. In short, you can use web-based media to showcase your administrations. The items you write about on your blog, in a creative manner will showcase your abilities and talents on the internet. Connect with web-based suppliers to expand your scope and use those abilities.

You can make sure to add all your web-based media and consumer accounts to your site so your customers can discover your data no matter where they are.

Think about blogging on your web. If you agree with the idea, you can use these web journals to prove your qualifications. Voters are more likely to respect your skills and give you a good score if you provide how-to posts, trends, and tips on your blog. To put this in its simplest terms, as a newcomer on the job, your activity across the Internet has as much impact as the material on your resume.

When all is in place, you will find clients who are willing to work with you. Don't worry about the large organizations; you have your work cut out for you. There are multiple tasks ahead of you. Do they think of trying out someone without a proven track record of success? You need to start with smaller organizations and hope to gain a few customers. Cold calling is one method you can use to sell yourself to people you expect would react favorably to your message. Cold mail is a method of communicating with someone with zero prior contact. For the situation I have in mind, LinkedIn would be better than email. LinkedIn is an expert group designed to help people establish personal connections and connect them with opportunities. People who hire creative candidates understand and accept the administrations that companies offer, explaining why they should do so.

Using a limited-time insight into your message will strengthen your appeal. Moreover, should they consider enlisting your help, they will spend less on time-tested executives or keep their proposal open.

Keep yourself up with web-based media trends. Your enemies must be able to catch up with you to stop you from winning. Regardless of that, to fall far behind in the market is to suffer a huge misfortune before the upswing arrives.

Most importantly, a company's web-level marketing needs to generate results. Thus, more consumers would be attracted. It is much more likely that more consumers will contribute to higher profits. As for that, however, your core value is, it ought to be your consistency. If you don't tell your customers anything, they will not be offended. Whether you need to begin with one client, you concentrate on high-quality rather than indiscriminate growth, and the rest will follow.

4. A Few Tips Of Note

- Whenever you've built up your web-based media showcasing organization, gotten a couple of customers, and made sensible progress, you will need to keep up any likely development. Leaving your office to deteriorate will mean it will fall behind different organizations and lose possible customers as time swims on, so you should be coordinated and predictable in your methodology.

- Whenever you've gotten a couple of customers and have been genuinely fruitful with them, consider adding an audit perspective to your site, connecting them with the surveys to help your believability for what's to come.
- Be available to change your technique now and then. Some of the time, a solitary methodology won't work in totally given situations. Then, you may be more adaptable to address your customer's and their clients' issues. Continuously be set up to examine how you can situate yourself to convey remarkable quality in each situation.
- Continuously research your crowd. Causing suppositions about what you to accept your crowd will appreciate, can be a hazardous game. Ensure that you utilize the entirety of the instruments accessible to you to discover however much as could be expected about your crowd socioeconomics in every informal community.
- Utilizing this information, focus on commitment. The degree of commitment you have decides how your crowd sees you and their eagerness to communicate. You need your crowd to feel like they have a functioning impact on the organization in your cooperation with them. Make a point to find some kind of harmony with them.
- You would prefer not to add an excessive amount of character that it takes from the special. In any case,

you would prefer not to be too special as this will drive away a portion of your crowd.

- Something you should take a gander, at for self-awareness is to define objectives for yourself. Set objectives that tackle your greatest difficulties. It may be to add greater character, be really captivating or be more influential in your promotion. This ensures you are truly improving and building up your mastery and profit your business over the long haul. Recall there is continually something new that can be learned. You ought to learn constantly.

There are likely a few more helpful hints and deceives. Nonetheless, I accept if this is your picked field for automated revenue, you will run over them during the time you spend mastering the abilities you need, just as during your time as a web-based media advertiser.

Until further notice, you have every vital stride to incorporate your way into the business and gain the pay you want. From that point, the capability of your business is just restricted by how far you will take it. Will you aim high? Or, on the other hand, will you try to achieve the impossible? The decision is yours.

Get Into Real-estate

The real estate market is likely quite possibly the most colossal business sector on the planet. Why? We as a whole need someplace to remain, or we need premises to go about as the base of our business tasks. Some portion of this market is the rental market. When a great many people can't or reluctant to buy a property, they may decide to lease, all things considered. While selling a property brings about the dealer accepting a mass installment at the worth of the property, rentals give you a fixed month-to-month payment at a charge you set for your property for some time.

Leasing offers you the chance to procure sufficient pay to take care of the expenses of that property that you need to fill and also give you extra. Obviously, there are sure dangers implied. You could wind up with a damaging inhabitant or a non-paying occupant. Notwithstanding, one could say that no type of automated revenue is without its dangers, and this is the same. You should do the perfect measure of examination into your likely clients before leaving the property.

While the real estate market is genuinely direct, there are a couple of interesting points to ensure you can capitalize on your business with decreased danger. If you have another property where you don't live or a storehouse appended to your home that isn't being used, at that point, a rental payment may be the correct route forward for you.

1. Renting Out Your Property

You need to consider your own property first when arranging a rental. What sort of property, right? Are your occupants going to utilize it for private purposes or business purposes? You should choose if the property is fit as a fiddle to be offered out. If not, you may have to put a little in fixing whatever is harmed or broken while making it look appealing, as could really be expected. Keep in mind, you need to discover inhabitants who will be cheerful living in your premises. This will mean they are bound to be solid and take care of their bills on schedule.

From that point, think about your area, the highlights of your property, and its

condition also. This will help you in deciding the cost of your rental. The overall guideline is that your rental expense is nearby around i% of your home's whole worth. In any case, note that a few spots may have power over your rental rate. New York, Maryland, and Washington all have differing levels of lease control laws, so you need to do some broad examination on the rental market and its wards around there.

2. Individual vs. Agency

When your property is fit as a fiddle, and you have thought of the amount you will lease it for, you can anticipate leasing your property.

There are two approaches to this. You could decide to employ an office that will deal with the business for you, or you could decide to do it without anyone's help. There isn't an excessive amount of mischief in doing it without anyone else's help, given you really understand what you are doing.

There are advantages to both. The rental organization offers you the advantage of having an attempted and tried technique regarding leasing properties to occupants. They will smooth out the cycle for you and handle all the desk work and correspondence from the occupant, handing it back to you.

You, for the most part, don't have a lot of anything to do with the interaction. The drawback is that specialists can charge a robust expense for their administrations, for the most part, 10% of the rental sum.

Then again, you could investigate the cycles and follow them yourself. In any case, a few occupants may have a conviction that it's simpler to exploit singular mortgage holders rather than the individuals who went through an organization. Should any issues happen, you may end up baffling with the methodology to follow. They will realize the immediate course to take on account of an organization and follow it to the tee.

3. Warning signs of being aware of

Anyway, for this situation, how about we simply accept you need to lease your property out freely. You should figure out how to vet your possible inhabitants. This is apparently quite possibly the main part of leasing. You need to examine a planned inhabitant's previous conduct as an acknowledged profile, just as their occupant profile confirms that they have been reliable and that there are no warnings to consider. Past conduct is normally a pointer of future conduct, so it's ideal to stay away from all customers who have any hiccups on their profiles.

There are a couple of dubious practices that you should know about going in. For instance, an inhabitant who requests to enter an understanding in an elective relative's name, outside of the individual answerable for the property, is an adequate danger to consider proceeding onward and tracking down an elective occupant.

Another warning would be a potential occupant who attempts to haggle on the underlying entirety to be paid. On the off chance that an occupant requests to pay the underlying charge in parts, this may show an absence of moderateness and will illustrate their future activities.

Try to assess your pet's approaches. Will pets be permitted on your property? On the off chance that you permit pets, what pets are appropriate to the kind of convenience you are giving.

At last, assess the sort of occupants you might want to have. More youthful inhabitants are bound to make hurt your property, just as occupants that have youngsters. Search for occupants who appear to be dependable and stable in their lives.

4. Rental Structure

When you feel that you have a solid handle on what to pay special mind to, you need to draw up a renting understanding ahead of time. This implies you are reasonably arranged whenever you have tracked down an occupant that is up to your guidelines.

Typically, it's feasible to discover a rent layout on the web; notwithstanding, it depends on you to guarantee that your rent has every one of the vital areas and subsections to fulfill your requirements.

For instance, a portion of the more broad parts of the rent to incorporate are:

- **Tenant Names.** This isn't only the name of the individual capable, however all occupants on the property.
- **Length of Tenancy.** All rental archives should state how long the term of the rent runs for. You could either have a tenant contract, which runs from one month to another and self-restores until ended, or a fixed-term rent, which typically keeps going about a year.
- **Rental Fee.** This ought to determine the expenses to be paid, when expected, and how it is to be paid. Consider including your installment technique just as any charges identified with late installments or check bobs.
- **Deposits and Additional Fees.** Be sure about what you may utilize the store for (harms, late bills, cleaning

expenses), just as what the occupant isn't permitted to utilize the store for (conclusive month's lease). It is likewise useful to incorporate where the store is being held and whether it will gather any revenue.

- **Damage and Maintenance.** Ensure the obligations of both the inhabitant and yourself are all around archived in this part with the goal that both you and the occupant know where they remain in such a manner. Remember any limitations for the occupant changing the property that you might want Pets. All limitations and conditions on pets should be set here.

When you are all-around set and coordinated, you can start to promote your property. Since there is a gigantic interest in the property, you will most unquestionably get offers for your property once it is out there. Putting commercials via online media and property sites, just as more discount sites will all carry imminent occupants directly to you.

As I said to start with, this is one of the more direct strategies for acquiring an automated revenue and perhaps the most solid. It can give a more steady balance to you to go out and live serenely.

Explore Cryptocurrency

There have been a lot of advancements in the digital money market recently. You may have learned bits and pieces of it here and there or pieces of it mentioned in the media.

If you'd put $100 into Bitcoin in 2011, you'd have over $1,000,000 today. At year-end 2017's end, bitcoin went up 1500% in value while Ethereum took off to over 10,000% in a single year. It's proof that has occurred of an enormously overinflated emotional reaction to digital currency. Although other cryptocurrencies have been volatile, there is no doubt that you will use cryptographic money to raise money.

You may be scratching your head trying to comprehend what I am describing here, so let me sum it up for you: digital currency

The aim of cryptographic money is to carry money in the same way that bookkeeping cash does. Cypher (informally, secret code) is the process of concealing data with cryptography. So far, no government or individuals have any direct or indirect influence over digital currency, which is a little weird. A record of transactions is held; for most parties, this is known as a blockchain. Now, a blockchain records permanent parts that can't be changed.

This record serves as an interchange. At the birth of cryptocurrency, more than 4,000 different kinds of cryptographic currencies have emerged. They're somewhat decentralized, and therefore, most of them charge their own fees and therefore aren't subject to expansion or depletion (Blockgeeks, 2019).

As the common wisdom goes, at the moment, people are unhappy with digital currencies. However, I think I am still a step ahead of the game as far as I am concerned. If the selected passage does not meet certain criteria, you cannot select a different passage from the data. At its most basic level, cryptographic money has no value.

It is possible to achieve success in digital money with a creative mindset and the correct degree of analysis.

1. Investing In Cryptocurrency

On the off chance that you've picked digital money to go about as your automated revenue, there will be a precarious expectation to learn and adapt you must dominate before you start. A broad examination into how the cycle functions and what steps you need to take to build up your payments are of the most extreme significance. Going in the visually impaired could see you losing all the worth you have reliably. Hence, it is significant that we limit hazards. Use test systems that work on the ongoing qualities to assist you with getting on the expectation to learn and adapt.

2. The necessary tools

Then, you will have to open up a couple of records, with the first being Coinbase. Coinbase will permit you to trade your money into digital currency for you to contribute with. You could consider utilizing another trade administration, anyway; this is the thing that I discover to be the most solid and secure.

From that point, you can set up a buy for whatever cryptographic forms of money you might want to buy, at the sum you need to contribute. Keep away from the more modest digital currency organizations, as you will be bound to succumb to tricks on any of those. Here, it is ideal to stay with the top suppliers on the lookout.

While choosing the amount you might want to contribute, I propose that your underlying venture be around 30% of your all-out speculation esteem. This means you should just contribute a little part of the aggregate sum you have with which to contribute, just to kick you off, and to guarantee you don't have a major misfortune. The measure of your underlying venture should be a sum that you are alright with making a misfortune on, should that happen.

Whenever you have bought your picked monetary forms, you will require a spot to protect them. This is known as a wallet. You can decide to either have a product wallet or an equipment wallet. Equipment wallets are utilized in a comparative way to outside hard drives. These can be utilized when you don't anticipate getting to monetary standards at any point shortly. Then again, programming wallets are utilized when you need to be dynamic in your exchange.

Make certain to expand your portfolio. As you would with ordinary ventures, don't toss all your investments tied up in one place. Expanding your portfolio will assist with safeguarding you when monetary standards become amazingly unstable and help to shield you from by and large misfortunes. Avoid versatile wallets, as these will be somewhat on the dubious side and are not as secure as programming or equipment wallets.

From that point, you're prepared to exchange. How you decide to exchange is simply dependent upon you and your system. You could decide to be truly forceful and go for transient gains or play the drawn-out game with your venture. The decision is yours.

3. A Bit Of A Heads Up

Set yourself up for this sort of market, should you decide to put resources into it. Cryptos are unreasonably unpredictable. An additional five minutes prior could transform into serious misfortunes an hour later. You need to prepare yourself for these events and guarantee that you don't carry on impulsive dynamics and feelings. This is the reason it is so imperative to just contribute however much you can bear to lose.

However, as I've effectively referenced, I will specify once more; stay with the more dependable digital money suppliers. Because of the troublesome and autonomous nature of cryptographic forms of money, it may be exceptionally simple to become involved with a trick. You should know what the signs that something isn't right are and when to move away.

Ensure that you have the motivation to enter each exchange. Exchanging hastily could see you losing at a consistent rate. Did you see an example? Something that caused you to consider purchasing in? Ensure that each venture has a purpose for it.

All things considered, try not to purchase basically because the cost is low. This is a slip-up a few beginners make. On the off chance that they see a drop in value, they will purchase in, accepting that it's less expensive and will ascend in esteem eventually.

At last, set objectives for yourself. It might sound somewhat unusual when managing a particularly unstable market, anyway defining objectives will assist you with setting up a mode of what you mean to accomplish. Given we're taking a gander at automated revenue, this could be an objective for you to arrive at every month. Nonetheless, for this situation, you should be adaptable in your objectives to oblige the unpredictability

Earn Through Google AdSense

If you consider, notices have been around for up to one can recall. They have been the main path for organizations to advance themselves and their item. As time has proceeded, promotions have remained, yet just the medium has changed. Presently you could likely discover advertisements in pretty much every spot possible, particularly in the online market. Pretty much every site has a couple of promotions to a great extent. Just a chosen handful of cases to be totally advertisement-free. This

could imply that, we huge loads of irregular promotions on shifting locales that hold no importance. That is the place where Google AdSense comes into play.
Presently this is something I referenced a little before while examining methods of making payments from publishing content to a blog website. Google AdSense shows advertisements on your site that are relatable to the substance you have, just as dependent on past client look. When clients begin clicking those advertisements, you will get installment as AdSense chips away at an arrangement of cost-per-click and shared income. This means you will need however many clients as would be prudent to tap on those promotions.

You will, by and large, get a commission cut going from anyplace between 20 pennies several dollars from these snaps. Keep in mind, this is procured at an expense for each snap, so producing a high pace of snaps could give you a significant expansion in income.

It is assessed that around 10 million sites are currently utilizing Google AdSense. They give security and straightforwardness to both the sponsors and distributors. Google ensures that the two sides have a reasonable comprehension of the entire cycle and everything can be followed.

On top of this, AdSense is fit for running an assortment of arrangements for its advertisements. You can run your advertisements in text, pictures, video, and a few others also. This allows experimentation to more readily suit your site. Google Adsense will likewise give you a choice to hinder advertisements that rival your substance. This is a convenient element when you track down that the promotions you are advancing are removing deals from you and giving them to your rivals.

1. **Using Google Adsense for returning money**

First of all, you will require a site or a blog. As we took a gander at already, you will have to zero in on a specialty you can base these advertisements on.

While applying for a Google Adsense account, there will be a couple of standards that you need to consider:

- **Ensure you own the Domain.** Part of the cycle of your application will expect you to check responsibility for space. Assuming you can't check that you own the space, Google will dismiss your application.
- **Quality Content.** We took a gander at this all through this book, so I envision it is nothing unexpected that to be thought of, you need to have quality substance. This doesn't mean you can grow a top-notch substance and tail off once you get a few promotions. Google will eliminate your record if they consider your substance to not be satisfactory.
  This is the reason it is imperative to think about energy as your hotspot for a specialty. You will be roused and headed to convey something fascinating

and significant on the off chance that you care about it.

- **The Age of your site.** This one is genuinely new from Google. You will presently be needed to possess your site and post substance for at any rate a half year before you apply for Adsense.
- **Google Policies and standards.** Ensure that your site clings to every one of the strategies that Google presents, just as their norms and limitations on a certain substance.

While this appears to be a great deal to consider, it is simply because Google needs to give the best substance compensation and guarantee that everything is reasonably directed.

Promotions are situated ideally to create more snaps. This will revolve around how you show promotions close by your substance. Ensure that they praise one another; your substance will, in any case, take inclination; however, the advertisements should, in any case, stick out.

Think about setting up your advertisements close by a Call to Action (CTA). A CTA is a procedure that trains clients to play out a specific activity on your site. Since CTAs are intended to cause them to notice themselves, setting advertisements around them might cause them to notice the promotions.

Presently we get to what I would think about somewhat of a hiccup. AdBlock instruments. As we discover promotions in more places around the web, more clients are going to AdBlock to confine advertisements from showing up, so they can zero in exclusively on content.

Lamentably, there isn't a lot of route around this. Everything you could do is basically amicably ask them to kill their AdBlock. An examination uncovered that more individuals would kill their AdBlock include if you essentially asked them to. Another thing to take into note is the likely measure of portable clients you may have. Cell phones are all over the place, and you would make a significant blunder on the off chance that you decided not to take into account this market.

You should change your site to get more versatile agreeable, yet you should change the promotions also to oblige that crowd.

2. Taking your income further

Suppose you feel that AdSense isn't sufficient for you in a perfect world. In that case, you could select to build your profit by consolidating with another type of automated revenue. Offshoot showcasing could be an extraordinary enhancement to the income you create from AdSense. You could likewise choose to sell your own items and administrations also. By the day's end, it truly depends on you how you decide to maintain your business. Albeit every one of these sections centers around one part of easy revenue, you could consolidate as numerous as you might want, as long as it is compelling and takes into account your way of life.

While Google AdSense appears to be somewhat on the stricter side, they are by and large requesting what most online suppliers request nowadays:

Consistency, appropriateness, and significance. Ensuring you are fully informed regarding their approaches and giving quality substance ought to, for the most part, be a standard on any stage you use, as I'm certain that you will become used to it instantly.

Google AdSense gives an extraordinary chance to procure pay while zeroing in your experience on other pay openings. Accept the open door!

Create Your Own Brand

High-creativity writing. For others, this would be confusing to a fault. To summarize, your picture is you doing whatever you can to get ahead. You're flaunting your new literary talents and skills for all to see. Your own image is one that individuals shape of you in an expert setting. It all depends on your presence on the internet and all things considered. Relationships are built by teamwork.

The majority of people are ready to talk only when they believe the individual they are dealing with shares an understanding. Each student gets his or her own label so that it's easier to associate and be credible with other students. Getting the right associations in the correct locations helps you become more grounded and makes you more accessible. You will be referred to as an expert in your field by others and be recognized for it. In one's own words, being allowed to design one's own works can lead to a sense of personal achievement.

Indeed, I understand. It does not increase sales. Considering everything, I'm here, Marking is a good strategy for drawing attention to the unique characteristics of the people who are involved with your brand. In all fairness, a company name gives you a larger platform to demonstrate your products or services. Think about it this way: If you already have an item on hand, bring in customers. Based on your current individuality, you now have an updated identity. Having an independent identity reveals how you are different from other specialists. Adoptions are good when they are in keeping with your personal brand image and when they can fuel your success and advancement over a longer period.

1. Creating Your Own Personal Brand

posting is the point at which you are composing an article and distributing it on another person's site or blog, permitting you to interface with a bigger crowd. This will assist with becoming you're after, and from that point, push on to more up-to-date adventures.

2. Host events and be a speaker at conferences

On the off chance that you need to fabricate your own image, you will have to comprehend that you need to put yourself out there, to be found in the public eye, regardless of whether via online media or as a general rule. Individuals don't work with organizations any longer. They work with individuals they like, given connections, associations, and trust.

Notwithstanding, it might be the beginning phases for you, and we need to begin at the nuts and bolts. Results will likewise not come straight away. Making an individual brand will take some time before you begin seeing any outcomes. When getting going, there are a couple of interesting points to get yourself going:

Blogging: Developing the positive image of your business by talking at meetings, maybe in any event and facilitating an occasion can spot you and your organization at the center of attention. The thought is that you need to utilize these chances to 'offer yourself to the crowd. Show them precisely why you are a specialist in your field and know the motivation behind your discourse. It might be ideal if you had somewhat comprehension of who your crowd is also. This will help you in having the option to draw in with them.

Crowd commitment is crucial, ensuring that your crowd is completely fixed on the thing you are saying. Think about an entertainer in front of an audience and how he covers the crowd in secret with his demonstration. You should accomplish something comparative, utilizing your words and the instruments you have available to you.

However, as of now referenced before, on account of individual marking, think about setting up your own blog. Be predictable and blog about subjects that are associated with your specialty. Offer your aptitude, and be intuitive with the crowd individuals you do get to guarantee they study you.

Ensure you know who your objective client is before continuing any further, as this will characterize who you contact.

3. Guest Posting

Whether you have your own blog, visitor posting can be an extraordinary method to acquire openness on more well-known stages.

4. Be active on social media

This is a given in this day and age. Ensure that you are dynamic on at any rate two informal organizations and that LinkedIn is one of them to draw in with more expert customers. There isn't a lot of pardon for not being dynamic via online media in the present time. It will bring about the vast majority thinking you have something to stow away. Regardless of whether that isn't the situation, individuals will be dubious of somebody who doesn't adjust to the set principles. It isn't for their premium that you will have web-based media, however, for you to advertise your own image and your business online on the most well-known stages.

The digital world is not the only world

Albeit the vast majority of the manners in which we have viewed at easy revenue all in all have been advanced, it isn't the best way to go about things. As you build up your own image, consider potential customers when you are making the rounds, consistently keeping business cards helpful, and dressing expertly. This will prepare sure that you're to consistently establish a decent connection, regardless of the situation. The individual brand is you. Thus you have to mirror that consistently. Your crowd is more trusting of what you may suggest. Guarantee that the items and administrations you advance are of the greatest quality before you advance them. Advancing and supporting low-quality items can be hindering your image and the trust the crowd has in you.

5. Monetizing Your Personal Brand

These are approaches to create and improve your own image. Clearly, to start, you need to have an idea, an arrangement, and a crowd of people you are attempting to reach. What we zeroed in on was developing that crowd. Presently you need to make money off of this difficult work you have been doing.

Here you should consider the different strategies we took a gander at past and a couple of impending techniques to develop your automated revenue.

- **Monetize your blog.** As an individual brand, you may not have any desire to simply litter your blog in advertisements for benefit. Think about courses and digital books as a way that will additionally improve your image and that you can look to start acquiring a pay.
- **Create an online community.** If you have an adequate crowd, you could think about opening up a private local area. Be that as it may, as a reference, you need to have enough clients for this local area to have the option to run all alone. For this local area, you can charge clients a month-to-month membership for access. If you're uncertain if this is a smart thought, inquire as to whether this is something they might want to see. Think of this way, you have another stage for a group of people to impart and develop from that point by assuming there's sufficient interest.
- **Affiliate marketing.** This is something we definitely think about. Your strong status as an individual brand will help your believability when showcasing items.

Develop An Mobile App/Game

In the current age, innovation is expanding at a dangerous rate. The interest for it is truly expanding also, with an ever-increasing number of individuals acquiring openness to the universe of innovation. This is particularly so on account of cell phones. As we progress, the cell phone is turning into a total need for anybody living in the cutting edge world.

With this interest in cell phones and handheld gadgets come the interest for applications that address each individual's issues. It is assessed that more than 5 billion individuals own a cell phone starting in 2019, so it's just characteristic that this market ought to be considered for forthcoming application engineers to get their hands into. Regardless of whether you're not an engineer, you could set aside the effort to become familiar with the exchange and think about making your own applications.

The business goes from singular designers to engineers making applications for huge organizations. From here, you have the chance to include yourself in an industry that is, as of now, encountering a development flood right now.

On the off chance that you want to tackle issues and some innovativeness to coordinate, at that point, this is simply the market you ought to consider getting into. For current designers, this is an awesome chance to transform your enthusiasm into something more prominent.

1. Creating Your First App

Suppose you have no related knowledge being developed. In that case, I believe that you realize the initial step is to get yourself familiar with the business. Begin learning perhaps the most well-known dialects utilized for versatility, like Java or Python. Work on making test applications, or in any event, cloning current ones as training. When you feel you're prepared to assemble your first application, we can proceed.

Conduct research. Check out other applications and see what separates them on the lookout.

- Consider their functionalities and what you could fuse into your own application to make it more fruitful.
- Consider the different ways each application brings in cash on their item, as you will require this data when building up your application later.
- Quest for holes in the market where there is a chance for you to grandstand an application.
- Try not to plan for an impressive future. From the start, your first application ought to be nothing excessively luxurious. Indeed, as you're new, it ought to be very fundamental.

This is only the way toward realizing the stuff to make your own applications. The solitary reason for your application until further notice is that it ought to tackle an issue. For what reason would it be advisable for it to tackle an issue? There will, at that point, be an interest in it. Once there is an interest in it, you can start to advertise it.

Research the laws and regulations. Before making your item, ensure it is by all laws and guidelines identified with a portable application. GDPR (General Data Protection Regulation), AML (Anti Money Laundering), and KYC (Know Your Customer) are largely fundamental with application advancement. You should be recognizable and proficient in every one of them, if your application is to be effective.

Begin planning your app. You should, as of now, have a thought of what you intend to make. Presently you need to consider what includes the application will have and what it will resemble. Conceptualize thoughts concerning the overall stream and design of that. Recall that your application ought to be outwardly engaging and easy to use. Likewise, you can design out how each component will chip away at your application and the cycles they will follow. This ought to incorporate the route of specific capacities and what will occur if a capacity is chosen.

Start building your app. When the arranging is finished, you will, at that point, need to begin assembling your application. To do that, you need the correct programming and apparatuses. Attempt to explore the best programming and instruments that will suit your necessities. JIRA, GameSalad, and Rollbar could be potential application creation programming that suits your necessities if you're not too keen on figuring out how to code. Be that as it may, these are by, and large extremely essential and are generally sufficient for the principal form of your application, yet nothing further. Appy Pie is an expected alternative if you are excited about composing the code for the application yourself. Everything relies upon what your requirements are and what you mean to accomplish.

Test your app. Almost every application needs to go through this testing stage. Application testing is there to ensure the application fills in as it ought to. There are no bugs or issues that will influence the client experience. Testing will likewise assist you with discovering approaches to make the application far superior. You need to have the option to build up the best application you can. Fixate and Jasmine are possibly acceptable application testing devices for you to begin with.

Release your app. Once every one of the bugs has been resolved and all that is by all accounts altogether, you need to put your application someplace that they can download and introduce it. The coherent decision is with versatile application stores. The two imposing business models in Apple and Google will need to get your item on their market. So how would we do this? All things considered, the cycles are somewhat unique.

Google Play Store. Join the Google Play Console and round out the different subtleties that layout your item. Google charges a one-time expense of $25 to make your engineer account. Free applications don't cost anything; anyway, Google takes 30% of the income for paid applications.

Apple Store. Apple costs are somewhat extraordinary as they charge a yearly expense of $99 to all engineers. With Apple, you should make an iOS dissemination provisioning profile and appropriation endorsement. You can do this through another of Apple's foundations called XCode. You will likewise have to set up your duty data just as pursue an iTunes Connect account. When this is finished, you would then be able to follow the strategy towards adding an application to the store. Your application will go for audit, and from that point, it's out of your hands for the present.

Monetization. If your application is a paid item, at that point, you've effectively adapted your item. Notwithstanding, you need to consider how you will do that on the off chance that you haven't. With portable applications, there are a few different ways you can conceivably make incomes towards an automated revenue:

In-app purchases. One of the numerous ways is through In-application buys (IAPs). IAPs are choices to buy an additional element to the application to make it a more charming encounter. These, for the most part, don't frustrate the general insight of the item, as they are typically only for added special visualizations.

Advertising. You could decide to show advertisements in your applications when clients are associated with the web. These promotions can show at specific stretches during the application. If you pick this course, ensure that your

application isn't over-burden with applications that ruin the client experience. You will likewise have to consider what kinds of advertisements you will utilize (pennants, show promotions, recordings).

Freemium products. These alternatives are very intriguing. The thought is that your item is accessible free of charge; anyway, free clients will have certain limitations for them that can be opened with paid substance. This is, for the most part, of the highlight of versatile games, where a client can acquire a serious edge from paying for their substance over different clients who decide to stay free clients.

Merchandise. Suppose you have items that compare with your application. In that case, you can consider promoting these on your application for any individual who might be keen on an expected reason.

Sponsorship. Albeit likely the hardest to achieve, sponsorships can be an incredible asset for creating income from your application.

This technique can likewise profit engineers as they acquire openness to a more prominent crowd.

There are a few alternative approaches to adapt your application too. Anyway, the above have commonly been demonstrated to be the best for most application designers.

2. Marketing And Promotion.

Presently that your application is finished, you need to construct your crowd for the application. This will require a touch of work, as application stores have many applications and yours is among those millions simply beginning. With such advertising, your application will conceivably go unseen.

You can do to battle this to guarantee that your application has incredible visuals that attract clients to it on the application store. Make a point to incorporate a definite and fascinating depiction alongside a name that sounds get and energizing.

Think about doing some broad publicizing of your item. This remembers contributing to promotions via online media, Google, and web journals too.

By and by, another chance could be for you to begin a blog devoted to your application. This can likewise give an extraordinary stage to clients to speak with you straightforwardly and construct a crowd of people for future applications.

Prior we referenced that it is alright for the principal form of your application to be an extremely essential variant. From that point, you should give refreshes that improve the nature of the application and its substance. On the off chance that your application stays as before constantly, individuals will gradually become exhausted and search for something better. It is dependent upon you to ensure your crowd stays drew in with your item.

Portable App Development is, without a doubt, a market where achievement can take you to the most mind-boggling statures, given its enormous interest. Indeed most achievement can be sufficient to construct an automated revenue that permits you to live uninhibitedly while performing something that satisfies you!

While there are many more ways to start earning and make a living, these ideas should give you a heads up as to where to go next.

Creating A YouTube Channel

Assuming you don't know about YouTube at this stage, you must have been sequestered underground for eternity. A behemoth of a rock.

Gone are the days when the allure of successful yet visually challenged visual substance was available on YouTube. Very correctly, YouTube is the largest video-sharing website on the internet. YouTube offers an endless supply of content from stories, video games to personal writings, trailers, films, teaching, first-person viewpoints, and creative viewpoint to teaching them to stories. For starters, clients can upload, purchase, and comment on each other's performances just like a new recording is added to a level. Since its conception in 2005, YouTube has gained 1.9 billion subscribers.

Even though YouTube is free, it does come with an additional option to subscribe to YouTube+, which offers additional benefits. YouTube Premium is a form of membership where clients are given advertisement-free substance, except for restricted content and out of blue onscreen use.

Video sharing is a common strategy for both self-publishing and independent clients and large corporations. Google permits YouTube and AdSense video ads to be considered "creative."

It's currently good to be a part of the YouTube fad, seeing how popular video blogging is right now. Discovering popular niche product fields that aren't overcrowded is difficult. Still, openings can be made even in most saturated markets if you've anything else to contribute. Regardless of whether you don't want to make YouTube videos for your company, using it for its advertising is sufficient. At this time, you will discover an excellent crowd of individuals who will be new to you on YouTube.

Lanze a de votre propre chaîne YouTube

Progressing company or only having to take care of your substance, you can never neglect substance. At that point, however, it is necessary to master the very basics.

Determine the purpose. You will have to know why you are making a channel. Is it to showcase your business? In case you're a substance maker, what substance do you wish to make? This will all become significant later when setting up your record. For possible substance makers, recall the significance of a specialty and an enthusiasm.

On YouTube, you can post recordings of pretty much anything. From computer

games to political sentiments, video blogs, and so on. Whatever is the energy you might want to share should be the issue here. Potentially the greatest illustration of this is PewDiePie. Beginning his YouTube divert in **2010,** Felix Kjellberg, or PewDiePie, as his username recommends, began his YouTube vocation making recordings of himself playing modest independent ghastliness games. He has proceeded to turn into the most bought-in YouTuber from that point forward from 2013 up until 2019 where he was overwhelmed. His channel moved into the runner-up, anyway, as yet figuring out how to get one of the lone channels to arrive at 100 million supporters. Everything began with enthusiasm...

Create an account The initial step you will have to make is to make a record. Odds are, you as of now have one on the off chance that you have a Google account. Notwithstanding, if you don't set one up so you can make your own profile on YouTube. Once you have a profile, you need to make a channel on YouTube.

Invest energy planning your channel, including channel craftsmanship to share your character and create revenue in your likely crowd. You could decide to transfer one of YouTube's formats as your channel craftsmanship or transfer your own that mixes your energy and interests into the channel.

You can likewise decide to add a channel trailer. At the point when somebody visits our channel, your trailer will help them discover what the issue here is. This video needs to catch the consideration of any individual who sees it and captivate them to look at the remainder of your channel later on.

The necessary equipment you have your channel set up; however, the thing would you say you will use to make recordings? At any rate, you will require the essentials, which are:

- **A webcam.** Nothing excessively extravagant. A quality top-notch webcam will accomplish for the present.
- **Microphone.** Ideally, settle on a commotion-dropping amplifier that records the sound independently. This will assist with forestalling any sound errors later on.
- **Screen Capture.** Suppose you anticipate basing your substance on computer games or instructional exercises. In that case, you will have to put resources into screen catch programming. There are free

choices accessible, like OBS Studio, that you can utilize.

Some other hardware to consider is a green screen. Assuming you need to change your experience, you will need to consider putting resources into a green screen with the goal that this is conceivable. If you don't have a webcam, and rather you have a camera or a cell phone, put resources into a stand to hold it steady when you are recording.

Create content Early on, you will require some substance that individuals can see on your channel. If you market a channel with no substance, this will just disappoint the possible crowd. To begin, a few recordings that are of value contained in your picked specialty. Around 1.13 ought to be an ideal add-up to keep a crowd of people occupied before all else while you begin to transfer more substance.

Promoting your channel Now that you have some substance, you can advance the channel. Recall that advancement is a drawn-out game. You should advance your substance for quite a while before acquiring sufficient fame to proceed with your development without it. So how might you advance yourself?

Social Media. Still the excellence of the advanced age. You can arrive at pretty much anybody via web-based media. This will allow you to fabricate a little base of devotees that will develop over the long run.

Work with other YouTubers. Suppose you can discover other YouTubers who will work together with you. In that case, this can be an incredible method to get new crowd individuals and extend your organizations.

Forums. Gatherings are the place where a great many people go to have conversations and visit about things. Consider looking for a local area that is pertinent to your substance. I would recommend being apart for a brief period.

Afterward, I respectfully propose that you've begun a YouTube channel and might want them to look at it. Along these lines, you can assemble compatibility with them first. From that point, you can utilize this to construct a more noteworthy parade of supporters.

Monetization You are continually going to have to discover approaches to build your procuring potential. Google AdSense will, as of now, be accessible to you; anyway, this should simply be an expansion to your principal source, as the income, you will create from AdSense won't be sufficient for you to have a nice easy revenue.

Affiliate Marketing will be one of the more famous decisions to associate with your YouTube channel. Suppose you've invested some energy in YouTube. In that case, you'll perceive the number of current YouTubers as of now use the subsidiary promoting to adapt their substance.

You ought to likewise think about selling stock. As you are presently a YouTuber, you will have many individuals who regard and appreciate the substance you make. Utilize this chance to sell stock that is revolved around your substance, which will help you turn out some significant items on the off chance that you are effective. At that point, there is the one that is turning out to be progressively well known today.

Become an influencer. On the off chance that you are an influencer in a specific specialty, you will begin being paid to elevate anything identified with that specialty. This can demonstrate a sound lift to your easy revenue.

Things You Need To Consider

As a free YouTuber, you will be liable for the entirety of your own substance. This implies that you should alter your own recordings to guarantee they are of the best quality. If you have Windows introduced, you may as of now have Windows Movie Maker. If not, consider financially savvy options that will coordinate with your necessities.

You ought to likewise design out your substance before you start. This guarantees that you have a sensible thought heretofore of what you want to do and how it is pertinent to your specialty. Bouncing straight in can leave you disarranged and lost, and you'll have squandered energy on a video you probably will not have any desire to transfer.

Focus on your watchers. If they leave remarks on your recordings, attempt to react to the most that you can, to show that you are perusing their remarks and that you're keen on what they need to say.

Be that as it may, disregard every one of the negative remarks. Try not to draw in with any type of antagonism that you find. For the most part, these are simply other people who seem to be terrible, and it would not look good for your picture.

YouTube, generally, is an extraordinary stage to impart your character and interests to the world while possibly producing a pay. With it, you can associate with a large number of similar individuals who are interested in your recordings and your character. Without a doubt, YouTube can be loads of fun!

CHAPTER FIVE

Researching and Studying the Market before a Business

So, now you know how to earn money and know about some stellar business ideas, and now you want to start your business, right? This is possibly going to be the trickiest part of your journey, and if you don't do this the right way, you might have a business that had failed even before it started! Hence, it is of paramount importance that you do thorough research of the market you aim to penetrate.

Initially, it might sound a bit daunting, but with proper and methodical planning, you will be able to make it work in no time. So far, we have been working at a very theoretical level, generating the ideas and deciding which business path we will pursue, right? This chapter will be the first chapter to actualize the vision that we have in mind, but, here's the thing, many young entrepreneurs don't understand the meaning of Vision or just how much weight it carries. Therefore, explaining what a Vision is will be our starting point.

Exploring Your Business Vision

A business vision is essentially a plan or idea you come up with to start a business. The majority of people who desire to get into business for themselves already have some form of the pre-existing idea of what they want to get into business with. This does not mean that you have the entire basis for the business planned out, but it does mean that you have a general idea of what you want to do.

The birth of a vision can come in number of ways. Some people are deeply passionate about something and desire to fill their schedule with this particular activity. They, therefore, want to start a business around it. Others see a need for a certain product or service and decide to create it and take advantage of this opportunity on the market. Furthermore, others may simply know that they want to go into business, and so they are looking for a vision to work with.

1. What is Your Vision?

If you already have a vision, now is a good time to get clear on what your vision is, exactly. The best way to do this is to write it down to get it out on paper. Consider everything that leads you to create this vision and everything that contributes to bringing it to life. For example, if you want to recreate an existing product to sell, explain why you want to do this and highlight how yours will be different from anyone else's that already exists.

Your vision is the inspiration that you have for your business. As you go about writing down your vision and the aspects that you already know, be sure to spend some time getting inspired and expanding your view. The more you know about your vision, the easier it will be to test it and turn it into an actionable plan later on.

2. What If You Don't Have One Yet?

If you don't already have a business vision, now is the time to come up with one. What exactly you end up with will be up to you, but you must choose something that you will genuinely be interested in. Picking a business vision that you have zero interest in will likely result in you losing interest in the business in general. When we don't pursue something that we are interested in or passionate about, we tend not to succeed.

That being said, begin exploring your options by considering what you are interested in and passionate about. See what comes up for you, and then go from there in researching business ideas that are available to you. The list of things that can be turned into a business is incredibly expansive. There truly is no limit on what can be done.

A great place to start your research is by going on the internet and researching businesses relating to your passion or interest. Then, go down the list of ideas you find until you come to something you genuinely care about and are interested in doing. From there, you can go ahead and proceed with developing your vision and turning it into a viable plan.

3. How Can I Develop my Vision?

Developing your vision will happen a lot more when you are in the process of turning it from a vision into a plan. For now, however, an effective thing to do is sit with the vision and brainstorm. Get really creative with this and take some time considering all areas. Write down any way you could be innovative and evolve your vision into something greater. This will help you immensely when it comes to turning it into a plan and outlining where and why you will be most successful.

At this time, there is no need to consider what is "reasonable." Play around with ideas and have an "anything goes" rule for what you come up with and write down. The more you flex your imagination now, the more unique your business idea may become. In many cases, this results in a business owner or an entrepreneur being left with an incredibly innovative idea that sets their business apart from others in a major way. Many business owners even experience massive success as a result!

And this is the point where the preliminary market research comes into play.

The Questioning Phase

Now, that you have your vision and idea down, it is time to begin discovering whether it can actually make you money or not. After all, you do not want to invest your time, energy, and resources into launching a business that will not generate you any income! In this chapter, you will discover how you can test your vision and ensure that it is one worth pursuing.

1. Is the Timing Right?

Time has a lot to learn. To be under a misconception is to many people, "There is nothing better than an idea whose time has arrived." Remember all of these things when you're getting your company up and running.

Make sure the world is ready to receive your idea before you launch. Many of those who went into business in the early 1990s did well because they included an online business model. Despite all this, though, the internet was still in its infancy. Exclusively online businesses will struggle due to lack of having insufficient traffic. However, there is an exception to this law, and that would be Amazon, which was founded by Jeff Bezos in the 1990s as an online-only business and is one of the most popular in the world today.

The majority of online businesses failed to realize a high return, but the ones who tried to learn something new. There was no chance of success until the owners were desperate to secure it.

It is true that nowadays, the online presence of a company is significant. I will admit, that is very subject to interpretation, as it depends on the type of company you're in. Even companies that have no web presence would likely profit from it regardless. since it's so easy and costs so little, this is considered one of the best tools for making new connections online. If you are not linked, you're missing out on one of the most important resources you have. This means you are getting less.

One of the main factors of any online business's success is good timing. Out of the starting gate, most could not keep up with the progress. For those who had the vision and anticipated it, nonetheless, they are living in the time of their lives, online and becoming successful entrepreneurs. They waited until the moment was ripe, raising their probability of success dramatically.

2. Are You Willing to Do Whatever It Takes?

As you read in the previous section, the exception to that wait-it-out internet rule was Jeff Bezos, who runs Amazon. This was not because he was lucky or did anything differently. In fact, everything he did essentially had the odds stacked against him. But, because he was willing to pursue it no matter what, he could generate a massively successful online company.

The same goes for you! No matter what you are seeking to pursue, you need to ensure that you have what it takes to make it successful. You should be willing to go all in. Business is not known to be easy. People who want it to be easy are employees, not business owners. Sure, many great freedoms and benefits come with being the owner. It's anything but easy, though, especially in the beginning when you are either completely alone or surrounded by only a limited number of other people sharing your vision.

If you want to know that your idea will be successful or not, consider what lengths you are willing to go to actually make it a success. Are you willing to pursue it at nearly all costs? Or are you only doing it because you want to make money and be an entrepreneur? Suppose you are not willing to suffer to some degree to make your business a success. In that case, there is a good chance that you will not make it over any of the hurdles you have to face as a business owner. And that's extremely important - especially early on - in the life of your business.

This does not mean that doing business is going to be some massive burden. In fact, it's typically quite the opposite. The majority of business owners will tell you that even though it can be an intense struggle and far more challenging than they ever anticipated, the journey of owning your own business can be deeply rewarding. This is even truer when you pursue something that you are genuinely passionate about.

3. Are Other People Actually Interested In It?

Finally, if you want to know that your business will take off, consider whether other people will actually be interested in it. As you probably know, a business without customers is no business at all. You need to pay attention to what other people are interested in. This will also help you determine the timing. Going back to the internet business industry: just twenty years ago, there was nearly no market there. This doesn't mean that no one was ever going to be interested in shopping online; the time simply had not come, and it was not something people were interested in *yet.*

Pay attention to other businesses and to trends that exist in society. Often, if you listen around, you can hear whether people would actually be interested in what you will offer. If the answer is no, you may want to hold on to that idea for a while and let it percolate. Or, you may wish to pursue something more solid and likable by the public in the meantime, letting the other idea sit on the back burner and wait for its time to come.

Turning your vision into an actionable plan requires a few steps. In this chapter, we will explore what is needed for you to transform your great idea into a business. If you're ready, let's begin!

The Research Phase

1. Research

The first step to turning your vision into an actionable plan is researching. Now that you have a brilliant idea that you have thought out, it is time to create something with it. You want to research everything from the best structure to pursue to your target audience. You will learn more about each of these elements throughout the rest of the book. Still, for now, you need to research enough to make a viable business plan. You will learn more about business plans in the next chapter.

Essentially, In order to have a clear idea about what you'll be doing or what plan you will execute, you would like to know everything that you can, about your business.

With diligent research and education, you can ensure that everything you do nurtures and supports your business growth. The more you know, the easier it is for you to take action and build your business.

2. Make a Timeline

No plan is successful without the timeline for you to create success with. You need to break your start-up goal into actionable steps so that you can ensure you're on track. What this will look like exactly will vary from business to business, but ideally, yours will look similar.

Set a date for the following parts of the process and ask these to yourself, plus any others that you decide will be necessary for starting up your unique business:

- When you want to have your business plan completed by?
- When you want to have enough start-up funds for you to begin?
- When you want to have your final product/service made by?
- When you want to have the structure for sales in place (i.e., rent a storefront, have your website set up, etc.)?
- When you want to start market research and testing your products/services?
- When you want to begin building the framework of your brand?
  When you want to start marketing your brand to your audience?
- When you want to officially begin selling to your customers?

These are all activities that need to happen to successfully launch your business with every element in place and ready to go. How long each of these steps will take may vary depending on your selling, how your business is structured, and more. This part of the planning stages can take anywhere from three months to a year. Some businesses find themselves continuing to nurture their start-up phase for a little over a year.

The best way to determine how long this will take you and make a reasonable and realistic timeline is to check out Chapters 7 to 9, where we discuss the basics of different businesses. This will help you get a better idea of what to expect. You can also use those chapters to help you determine everything that needs to go on your timeline. Lastly, you might consider doing some internet research on what other companies similar to yours have done and how long their start-ups took

3. Take Action

Finally, executing the strategy is crucial! Before implementation, double-check your action plan so that you're still aware of its status. If you have any time available to spare, spend some time working your way through the timeline at least a few days per week or every other week. It's the perfect balance of structure and flexibility to give you the maximum amount of time to devote to your startup.

In the minds of many entrepreneurs, the startup phase has already slipped their attention. When you're down on your money, your focus or effort will suffer. If your concept was not important enough to merit the attention you paid to it, you shouldn't have taken it to market in the first place. If that's the case, then you'll have to show some self-control now. Work hard, be serious, and keep going.

When you don't feel like giving a hundred percent effort, you can sign up for 100 percent effort. Don't worry about him - this person doesn't affect your company, but who's a cheerleader when you need it helps you stay on track. This is to say that you give them the same motivation and support. This is a perfect way to remain focused, as well as open up your possibilities and sights. Starting a company is a great way to forge many personal friendships as well. You may want to seek out a partner to help you stay focused.

If you prefer, you might try to find a teacher. Business advisors are people who have successfully or are currently running a similar business to yours. Your mentor should have the same beliefs and abilities that you possess. A mentor can help you get your questions answered and provide advice, or they can assist you in putting those ideas into action and enabling you to succeed in the early stages of starting your own company.

Support is important in every respect. If you feel like you need a little extra help or additional assistance, know that you have plenty of opportunities to choose from. Ask for support when you need it, but don't depend on it.

That's about right. Numerous studies have shown that the most troublesome point for novice company owners is the subsequent method of conducting market research.

Learn the rest of the chapter about proper market research, and we will explain how you can go about it.

What Is Market Research?

Market research is described as determining the viability of a new product or service via direct consumer research. This approach enables organizations or companies to identify their target group, gather and record feedback, and make informed decisions.

Market research can be performed directly by organizations, businesses or outsourced to agencies with experience in this field. Market research may be accomplished by deploying surveys, communicating with a community of people known as a sample, performing interviews, and other similar methods.

The primary goal of market research is to understand or analyze the market associated with a specific product or service to determine how the consumer will respond to a product or service. The knowledge gleaned from market research can target marketing/advertising campaigns or ascertain what consumers' function priorities/service requirements (if any) are.

1. The Key Objectives Of Market Research

Typically, a market research project would have three distinct forms of goals.

- **Administrative:** Assist in the growth of a corporation or industry by proper planning, coordination, and management of both human and material capital, and thereby meet all relevant consumer needs at the right time.

- **Social:** Meet a customer's individual needs by providing a necessary product or service. When a product or service is consumed, it should meet the needs and desires of the consumer.
- **Economical:** Determine the economic degree of success or failure a business may have when entering a new market or launching new goods or services, offering certainty to any steps to be taken.

The Importance Of Market Research

Conducting research is one of the most effective methods of increasing consumer loyalty, decreasing customer turnover, and elevating the market. The following are some of the reasons why market research is relevant and should be considered in every business:

- It offers valuable insight and opportunities about the potential of existing and new goods, allowing companies to prepare and strategize accordingly.
- **Customer-centric:** It aids in determining what consumers require and want. Marketing is centered around the consumer. Knowing the customer and their needs can assist companies in designing goods or services that better serve them.

- **Forecasts:** By understanding consumer expectations, companies can predict their production and sales. The market research also aids in deciding the best inventory stock to keep on hand.
- To maintain a competitive edge, market research is an important method for conducting comparative studies. Businesses should develop business strategies to stay ahead of their rivals.

The Different Methods Of Market Research

Market research assists organizations and businesses in drawing concrete conclusions about consumer purchasing habits or the probability of customers paying a certain price for a commodity.

The following forms are available depending on the methods and resources required:

- Primary Market Analysis (a mix of qualitative and quantitative data
- Secondary Market Analysis

1. **Primary Market Analysis**

Primary market analysis is when companies or businesses contact end-users or hire a third party to conduct relevant surveys to collect data. The information gathered may be qualitative (non-numerical) or quantitative (numerical) (numerical or statistical data).When performing primary market research, two types of information can be gathered: **Exploratory and Specific**. Exploratory research is a type of open-ended research in which a topic is investigated by asking open-ended questions in a detailed interview format, typically with a small group of people known as a sample. The sample size is limited to 6-10 people in this case. On the other hand, specific research is more oriented and used to solve problems found by exploratory research.

As previously said, primary market research is a mixture of qualitative and quantitative market research. A qualitative market analysis study collects semi-structured or unstructured data using some of the most widely used qualitative research techniques, such as:

Focus groups: One of the most popular qualitative analysis approaches is the focus group. A focus group is a small group of people (6-10) who respond to online surveys sent to them. The best thing about focus groups is that the information can be gathered online without communicating with the group members in person. However, since it is used to gather complex information, it is a more costly tool.

One-on-one interview: As the name implies, this approach requires personal contact in the form of an interview. The researcher asks a series of questions to the respondents to gather information or data. The questions are often open-ended and posed in a way that encourages answers.

This approach is highly reliant on the interviewer's skill and expertise in asking questions that elicit responses.

Ethnographic research: This form of in-depth research is performed in the respondents' natural settings. This approach necessitates the interviewer adapting to the respondents' natural environment, which may be a city or a remote village. Geographical restrictions may be an impediment to performing this type of study. Ethnographic research will last anywhere from a few days to several years.

Organizations use qualitative research techniques to perform systematic market research using online surveys, questionnaires, and polls to obtain statistical information to make informed decisions.

Previously, this process was carried out with a pen and paper. This has now progressed to the distribution of organized online surveys to respondents to obtain actionable insights. Researchers typically organize and construct their surveys using current and technology-oriented survey platforms to elicit the most responses from respondents.

Data is quickly gathered and recorded using a well-structured mechanism. Necessary action can be taken from all of the knowledge that is made available firsthand.

2. **Secondary Market Analysis**

Secondary market research makes use of knowledge organized by outside sources such as government departments, media, chambers of commerce, and so on. This knowledge is widely disseminated through newspapers, magazines, books, business websites, free government and non-government organizations, and so on. The following are examples of secondary sources:

Public sources: Such as libraries, are fantastic places to get free information. Government libraries typically have free resources, and a researcher may record available information.

Commercial sources: Commercial sources, while accurate, are often costly. Local newspapers, magazines, journals, and television are excellent commercial sources of knowledge.

Educational Institutions: While not a common source of information, most universities and educational institutions are rich sources of information because they conduct more research projects than any other business field.

The Core Steps To A Market Research

Knowing what to do in different scenarios during the investigation will save the researcher time and help minimize issues. Today's successful businesses employ powerful market research survey software, enabling them to perform extensive research on a single platform and thereby provide actionable insights much faster and with fewer issues.

The steps for conducting successful market research are as follows.

Step 1: Identify the Issue

Researchers can benefit from having a well-defined research topic as they ask questions. These questions should be geared toward problem-solving and should be tailored to the project. Ascertain that the questions are written explicitly and that the respondents comprehend them. Researchers should run a test with a small group to see if the questions are understandable and sufficient to yield informative results.

Research goals should be written precisely and provide a brief explanation of the information required and how it will be obtained. They should be able to address the question, "Why are we doing research?"

Step 2: Specify the Sample

Researchers need a representative sample to conduct market analysis, which can be obtained using many sampling techniques. A representative sample is a small group of people that correctly represent a larger group.

A company cannot afford to spend money gathering knowledge from the wrong population. The population in the sample must reflect characteristics that are significant to the researchers and that they need to investigate.

Consider that advertisers will always be vulnerable to bias in the sample because there will always be people who do not answer the survey because they are too busy or answer it incompletely, resulting in researchers not obtaining the necessary data.

In terms of sample size, the bigger the sample, the more likely it is to represent the population. A larger representative sample gives the researcher more confidence that the people involved are the ones they need, minimizing bias. As a result, if they want to avoid inconsistencies in our surveys, they should use representative and balanced samples.

Almost all surveys that are taken seriously are focused on scientific sampling, which is based on mathematical and probability theories.

A representative sample can be obtained in two ways:

- Probability sampling: In probability sampling, the sample is chosen randomly, ensuring that each member of the population has the same chance of being selected and included in the sample community. To ensure representativeness, researchers should ensure that they have up-to-date information on the population. They will draw the sample and survey the majority.

- Non-probability sampling: In non-probability sampling, various groups of people pursue a more balanced representative sample. Knowing our group's demographic characteristics would certainly aid in narrowing the profile of the desired sample and defining the variables that concern the researchers, such as gender, age, place of residence, and so on. Researchers can monitor creating a representative sample that is efficient for us by understanding these requirements before obtaining the details. There may be a margin of error where a sample is not representative. If researchers want a representative sample of 100 workers, they should choose a balanced mix of men and women.

While the sample size is critical, it does not guarantee accuracy. More than size, representativeness is linked to the sampling frame or the list from which people are drawn, as in a survey.

If researchers want to learn more about determining sample size, check out our sampling guide here.

Step 3: Conduct data collection

To begin, a data collection instrument should be developed. They can cause errors in research if they do not respond to a survey or answer it incorrectly. This can be avoided if data is collected correctly.

Step 4: Examine the findings

Each stage of the market research phase is related to the next. If all of the above is done correctly, but there is no accurate interpretation of the outcome, the decisions taken as a result would be incorrect. In-depth research that is carried out without leaving any loose ends would be useful in obtaining solutions. The results of the data review will be recorded in a study, which should be written explicitly so that successful decisions can be taken on that basis.

Analyze and interpret the findings in the process of looking for a broader context in the collected data. To get to this stage, all of the previous steps had to be completed.

How do researchers assess the outcomes? The only quantitative details collected are age, gender, occupation, and the number of interviewees because the rest are feelings and perceptions that the interlocutors have relayed to us. There is a tool called an empathy map that forces us to put ourselves in our customers' shoes to recognize the characteristics that will enable us to create a better fit between our goods or services and their needs or interests.

When the study is carefully designed, the hypotheses are properly established. The appropriate selection method is used; interpretation is typically simple and accurate. What happens next after performing market research?

Step 5: Write the Research Report

When presenting the findings, researchers should consider what they want to accomplish with this research study. They should not conclude that the survey framework is the best way to conduct the review. One of the most common errors that many researchers make is that they present their findings in the same order as their questions and fail to recognize the power of storytelling.

The best analysts advise on how to write strong reports: To present the findings, use an inverted pyramid format, starting with the key questions of the company that prompted the investigation.

Instead of collecting facts, start with the assumptions and give them fundamentals. Following that, researchers would be able to provide information to readers who have the time and interest.

Step 6: Make a Choice

A company or analyst should never inquire, "Why do market research?" They should simply do it!

A market research study may provide analysts with a wide range of details, such as customer buying intentions or input on the growth of the target market. They will also find useful information that will assist them in calculating the prices of their product or service and determining a point of equilibrium that will favor both them and the customers. Make a choice! Act and put your ideas into action.

So now, you've put all the effort into doing your research, what would be your benefit from all of these?

The Benefits Of Market Research

Make well-informed decisions: The growth of a company is based on how management makes decisions. Using market analysis methods, management can make business decisions based on data that supports their expertise and experience. Market research helps to understand market dynamics, so it should be done regularly to thoroughly know the customers.

Accurate information: Market analysis offers true and accurate information to help the company plan for any potential mishaps. By thoroughly researching the market, a company will inevitably take a step forward. As a result, it will gain an advantage over its current rivals.

Calculate the market size: A researcher should assess the size of the market that must be protected to profitably sell a product or service.

Select a suitable sales system: Choose a precise sales system based on what the consumer requires, and the product/service can then be placed in the market.

Learn regarding consumer needs: It is beneficial to understand how clients' preferences (and tastes) change to meet preferences, buying patterns, and income levels. Based on the particular needs of customers, researchers will decide the type of product that must be produced or sold.

Collect information about the brand's perception by customers: Market research, in addition to generating data, assists a researcher in determining how consumers view the company or brand.

Analyze consumer engagement methods: Market analysis can be used to direct communication with existing and future customers.

Productive business investment: It is a great investment for every business. It provides useful knowledge and shows researchers the direction to take the right path and achieve the necessary sales.

CHAPTER SIX

Creating the Perfect Business Plan

Now that you have come up with a great business idea and validated it, the next step is to create a business plan. This is a very critical step if you want your business to succeed. This is where you get to plan how you will execute your business idea. As the saying goes, failing to span is planning to fail.

Despite being a crucial document, many would-be entrepreneurs find it very challenging to come up with a business plan, with some even choosing to start their business without one. Don't be tempted to take that route.

In this chapter, I will teach you everything you need to know about business plans, including what a business plan is, why you need one, and how to write a great one.

I will also give you a template that you can use as a foundation to write your own business plan.

What Is A Business Plan?

Like you can probably infer from the two words, a business plan is simply a plan for your business. It is a document that acts as a roadmap for your business. It describes how your business is going to work, outlines of the goals and objectives of your business, and the detailed plans that will help your business achieve these goals and objectives. In a nutshell, it is simply a guide on how you will make your business succeed.

Why Do You Even Need A Business Plan?

There are several reasons why you need to come up with a business plan before launching your business. Some of these include:

1. **To Secure Funding**

In most cases, new businesses will require some startup capital to get things rolling. Sometimes, however, the business owner might not have the money required to start the business. In such cases, the entrepreneur will have to rely on debt or equity financing from financiers and investors.

With the high rate of business failure, investors and financiers want to be sure they are putting their money into a venture that will be profitable, not something that will go under soon after, and drag down their money with it. The business plan provides them with a tool to evaluate the feasibility of the business before they decide to invest in it.

Without a business plan, your chances of securing funds to start your new business are next to nil. Therefore, if you consider borrowing some money to fund your new business, you need to develop a business plan.

2. For Planning Purposes

This is the main purpose of a business plan. Creating a business plan helps you think about everything required to make your business a success. It will help you plan the best way to solve your customers' problem, how to promote and market your products and services, how to win market share from the competition, the time you need to get to certain milestones, the resources you need to get your business off the ground, and so on.

As you write your business plan, you will also identify potential problems you might encounter. All this will make your startup journey a lot smoother and increase your chances of success. A 2010 study on business planning by Andrew Burke, Stuart Fraser, and Francis Greene, published in the *Journal of Management Studies,* found that companies with business plans grow 30% faster than those without.

3. To Know If Your Idea is Feasible

This is perhaps the most popular reason (but not the most important) behind writing a business plan. Just because you have a profitable business idea doesn't mean that you will achieve success. It all depends on how you execute the idea. To know if your execution plan is feasible, you have two options. The first one is to launch the business and hope everything works. This is a very expensive and time-consuming way of finding the feasibility of your idea. The second option is to create a business plan to test the feasibility of your execution plan. The process of creating a business plan will reveal whether or not there are any flaws in your execution plan. This will save you a lot of time and money, which you would have lost if you launched without a business plan.

4. For Research Purposes

To come up with a business plan, you will need to conduct lots of research. Research about your ideal customer, research about the size of the market and market trends, research about your competitors, and so on.

This information is very important since it will help you develop strategies to ensure your business's success.

5. Ease The Partnership Process

Suppose you decide to approach other businesses for collaboration and partnerships. In that case, both of you need to be on the same page before embarking on a partnership. Your business plan will provide your potential partner with a dear overview of your business objectives, your audience, your vision, and your strategies for growth.

This information will be useful in helping your potential partner to determine if your business is a good fit for them, especially if you are trying to partner with a bigger business than yours.

6. To counterbalance your emotions

Running a business is an emotionally charged affair. Sometimes, your passion for your business and your ideas might drive you to make unrealistic assumptions. Sometimes, you might be overwhelmed with fear or self-doubt. In some cases, these strong emotions might get the best of you, resulting in rash decisions that could be detrimental to your business. However, when you have a business plan that outlines what you need to do and for what purpose, you are less likely to make irrational and emotion-based decisions.

Steps To Writing/Drawing A Business Plan

Contrary to popular belief, drawing up a business plan is quite easy. However, you will need to dedicate time and effort to carry out the needed research for various sections of your business plan. While there are different kinds of business plans, depending on the nature of the business you intend to start, some key elements are found in most business plans.

1. The Executive Summary

This is usually the first part of a business plan. The executive summary provides a quick overview of what is contained in the rest of the business plan. It gives an outline of your business, what you do, who you are targeting, and some key financial highlights.

The executive summary aims to give readers without a lot of time (such as potential investors) the big picture of your business. Ideally, the executive summary should be able to stand on its own. It should be very clear and concise, highlighting all the key aspects of your business without going into too much detail. You can think of it as a one-page business plan. Some investors will only ask for your executive summary rather than the whole business plan.

Suppose they like the information presented in the executive summary. In that case, they might then ask for something to shed more light on your business, such as a pitch presentation, a complete business plan, or more detailed financial information. Therefore, you need to make sure that your executive summary is intriguing and compelling.

However, if your business does not require financing or partnerships and you are only writing a business plan for your own purposes, you can leave out the executive summary.

Even though the executive summary is the first chapter of your business plan, it is recommended that you write it last after you have written the rest of the business plan. This way, you will have a dear idea of what is contained in the rest of the business plan. Remember, the executive summary is, just as it sounds, a summary of your business plan.

Ideally, your executive summary should be about one page in length. Below are some of the things that the executive summary should cover:

Problem Summary: An overview of the problem you are trying to solve for your customers. This is basically a justification for why your business needs to exist.

Solution Summary: An overview of your business, its products and services, and how they are different from what is in the market.

Target Market: A brief overview of the market segment you are targeting.

Competition: An overview of your key competition.

Marketing Plan: A brief summary of how you intend to reach your

customers.

Current Financial State: A brief overview of your company's current financial status.

Financial Forecasts: A brief overview of the financial targets you intend to hit in the next 6 months, I year, or 2 years.

Financing Needed: If your business plan aims to secure funding, give a brief overview of the amount of money you need. If you aren't trying to secure funding, leave out this part.

Why Us: A brief overview of the people involved in running the business and why they are the right people to run this business.

2. The Opportunity

This is the first chapter of a business plan. This is where you go into detail about the problem you are trying to solve, how you are solving it, the market segment you are targeting, how your product fits into the market, who your competition is and how you will deal with them, and so on. This chapter should contain the following sections:

The Problem Worth Solving: In this section, you will define the problem that your business will solve for its customers. Here, you will need to answer Questions such as what your customers' primary pain point is? What are the Existing solutions for this pain point? Why do you feel that the existing Solutions are not effective? To be able to properly define the problem, you will Need to go out there and talk to potential customers. What you think is there The greatest problem might actually not be a problem. The only way to validate Your assumption is to talk to them.

The Solution: In this section, you should go into how you intend to solve the problem you identified above. What are your products or services? How do they solve the problem? Why are they better at solving the problem compared to what is in the market?

Target Market: After defining the problem and describing your solution, the next thing is to explain who you intend to sell to. Here, avoid the temptation to describe your target market as everyone. Just because you have a product that everyone can potentially use doesn't mean that everyone will be willing to buy it. Identify the group of people for whom the problem you are trying to solve is a real pain point. Having a good idea of who your target market is will help you come up with the right sales processes and marketing campaigns.

Market Research: Don't just stop at describing who your target market is. Go a step further and conduct research into how big the market is and the different segments within that market. This will help you determine if the market is big enough to support the success of your business.

Market Trends: In this section, you will describe the notable changes that are happening in your target market. Are there any significant changes in the needs and preferences of your potential customers? Is there a new market segment starting to gain interest in the kind of products or services you are offering? How do you intend to take advantage of these changes to give your business an edge?

Market Growth: This section will explain the growth your target market has undergone over the last couple of years. Is it growing or shrinking? If the market is growing, this is a good sign since it shows an increasing demand for solutions to the problem you are trying to solve. If the market is shrinking, this is not a very encouraging sign. However, it is still possible to succeed in such a market. However, you will need to understand that you are swimming against the current. To find out the kind of growth happening in your target market, you will need to carry out a lot of research.

Competition: In this section, you will describe who your key competitors are. To determine who your competitors are, you should think of the various ways through which your customers currently solve the problem you are trying to solve. Here, even businesses that do not seem to be direct competitors could still compete with you. For instance, if you are launching a beer brand, you might assume that your only competition is other beer brands and alcoholic drinks.

However, if someone is sitting in a restaurant waiting for someone, a question like this might go through their mind: should I drink a beer or soda? Because both your beer and soda are solving the same problem (the customer needs something to drink), they are both in competition, albeit indirectly. Apart from describing your competition, you should also describe which competitive advantages you have that will help you gain a chunk of the market share. Do you have access to new technology that your competitors do not? Do you have some intellectual property rights that make your product or service superior?

3. Execution

This is the second chapter of your business plan. This is where you will go into the details about how you will make your business work. In this chapter, you will cover things such as how you will sell and market your products, how your business will operate, how you will measure success, and so on. This chapter will include the following sections:

Marketing Plan: Here, you will describe the strategies that you will employ to reach your target market and how you intend to position your business and your products and services in the market. Your chosen strategies will depend on your ideal customer. For instance, if your ideal customer spends more time on Instagram than Facebook, then part of your marketing plan will include Instagram marketing. Your marketing plan should include details such as your pricing plan and the reason why you decided to price your products that way, your positioning statement, which explains how you will position your products and company to differentiate yourself in the market, and your promotion strategies, including the reasons why these strategies are the best suited for your target customers.

Sales Plan: This will describe how you will get your customers to buy your products and services. Mention any sales channels and processes you have in place, the people who will be in charge of selling your products, as well as the place where you will sell your products.

Operations: This section details how your business will operate. Here, you will need to describe things such as the people or companies that will supply you with the raw materials you require for production, the locations and facilities where your products will be produced, and how you will get the Products to the customers. If you are not manufacturing your own products, describe how you will get the products, whether this is through partnering With wholesalers or through Dropshipping. You should also describe all matters to do with your inventory, such as how much inventory you intend to maintain, where this inventory will be stored, how you will monitor and track inventory, and how you will deal with unexpected spikes in demand and busy Seasons. If you have some technology that will play a key role in the operation of your business, describe it and how it works.

The point here is to show the reader that you have a pretty clear idea of how your business will work and of your supply chain, and that you have solid plans in place to deal with Potential uncertainties. If you are writing the business plan for yourself, the operations plan will help you make key decisions about your business, such as how to reduce costs, how to price your products, how to plan your break-Even point, and so on.

Milestones And Metrics: For a plan to be useful, you need to break it into steps that you need to follow while implementing the plan. In this section, you will describe the steps you need to take, the dates before which the steps need to be completed, and the people responsible for each of these steps. In addition, you will also describe the key metrics that you will use to monitor and keep track of how your business is growing. Examples of metrics that you might want to monitor include the number of leads generated, sales, website visitors, or any other metric that will give you a good idea of how your business is growing. The milestones and metrics section does not need to be overly long. The point is to make it clear what steps need to be taken to make your business successful.

Key Assumptions And Risks: Here, you are going to describe the key assumptions you have made that are crucial for the success of your business, Both favorable and unfavorable. For instance, if the success of your business depends heavily on Instagram marketing, you are making the assumption. That Instagram will always be a great platform for marketing.

If Instagram were to shut a few months down the line, this would pose a huge threat To your business. The good thing with knowing your key assumptions and Risks are that it allows you to prepare. If you know your key assumptions, you will focus on proving if they are correct. The aim is to minimize assumptions. Therefore, remove any uncertainties from your business.

Similarly, when you know the risks you might face, you will start developing contingencies before these risks become actual problems.

4. The Company

This is the third chapter of your business plan. This chapter provides an overview of your company, including the structure and the people involved with running it. If you are writing the business plan for your own use, you have the option of leaving out this chapter. It is usually the shortest chapter in the business plan, and consists of the following sections:

Company Overview: In this section, you are going to describe the ownership of the company, its legal structure (sole proprietorship, incorporated company, limited partnership, or general partnership), a brief history of the company, and why it was founded as well as its location.

You will also include information such as the mission and vision, the company's values, any intellectual property held by the company, the nature of the business, the industry in which the company is operating, and your business objectives.

The Team: In this section, you will give a list of all the key members involved in the running of the business, a brief bio of these team members, as well as an explanation as to why these people are the right people for these roles. Most businesses succeed or fail to depend on those who are tasked with running them.

Moreover, investors will want to know the people behind the company and their qualifications before putting money into the business.

Even if you are writing the business plan for your own use, this section can help you discover any oversight in the team tasked with running the business.

5. Financial Plan:

This is the final chapter of your business plan. This chapter is crucial because running a successful business involves paying close attention to your income and expenditure. Having a solid financial plan will help you make key decisions like deciding when you can purchase new equipment, bring in new employees, expand the business, and so on.

Suppose you are trying to secure funding from investors or financiers. In that case, a good financial plan will also help you determine how much money you need to get your business off the ground, which will, in turn, help you determine how much money to borrow.

This chapter consists of the following sections:

Sales Forecast: In this section, you are going to provide projections or estimates about the number of sales you expect to make over the next two or three years. You don't have to go into too much detail at this point. Just provide a high-level overview of how much you expect to sell. If your business has multiple products or services, you should provide projections for each product or service separately.

Personnel Plan: In this section, you will go into the details of employee salaries. If you are starting a small business, you can create a list of all the people you intend to hire and the amount of money you will pay each of them every month. For a big company, this might not be feasible. What needs to be done in this case is a breakdown of the personnel plan by departments, such as "sales," "finance," and "IT." You are also going to include the employee burden in this section. Employee burden refers to other costs that come with having employees on your payroll other than salary. This might include things such as insurance and taxes.

Profit And Loss Statement: Also referred to as an income statement, this is a document that brings together all your numbers and helps you determine whether you are making or losing money. The income statement includes all your sources of revenue within a given period (extracted from your sales forecast) and your expenses over the same period (personnel costs, production costs, and all other expected expenses). The income statement gives a bottom line by subtracting the expenses from the revenue. It helps you determine whether you are going to be making profits or incurring losses.

Cash Flow Statement: The cash flow statement is almost similar to the income statement and sometimes even gets confused with the income statement. Like the income statement, the cash flow statement consists of a list of your revenue and expenses.

However, there is one key difference between the two. Whereas the income statement will include revenue from all sales within a certain period and all expenses from the same period, the cash flow statement also considers when your company collects revenue and pays out its expenses. As a result, it is a great tool for keeping track of the amount of cash that the business has at hand at a certain time. Unlike the income statement, the cash flow statement does not calculate profit and loss. By keeping track of the amount of cash you have at hand, the cash flow statement helps you determine when you have surplus money or low amounts of cash and might need to borrow money to keep your business running. Since, your business is still new and money has not yet started flowing, you should create a projected cash flow Statement for the next twelve months.

Balance Sheet: This is a financial statement that gives you a look into your company's financial health. The balance sheet consists of a list of all the assets owned by your business (equity) and a list of what you owe (liabilities). Subtracting the total liabilities from total assets will help you determine how much your business is worth.

That pretty much covers the basics of what a Business Plan is and how you can prepare it. I have provided a template in this chapter that you can use as a point of reference when preparing your own business plan.

Sample Template

To make it easier for you to create a business plan, below is a template that you can use as a foundation.

BUSINESS PLAN[DATE]

Contact Information

[Telephone]

[Email]

[Website]

[Address]

TABLE OF CONTENTS

The Financial Plan

Sales Forecast

Personal Plan

Profit and loss statement

Cash Flow Statement

Balance Sheet

Executive summary:

Target market

Competition

Marketing Plan

Current Financial state

Financial Forecast

Financing needed

Why us

CHAPTER ONE: OPPORTUNITY:

Problem Worth Solving

Market Research

CHAPTER TWO: EXECUTION:

Marketing Plan

Sales Plan

Operations

Milestones And Metrics

Key Assumptions And Risks

CHAPTER THREE: THE COMPANY:

The Company Overview

The Team

CHAPTER FOUR: FINANCIAL PLAN:

Sales Forecast

[insert tables or charts to show sales projections]

Personnel Plan

[insert tables and charts to show salaries and employee burden]

Profit And Loss Statement

[insert profit and loss statement]

Cash Flow Statement

[insert cash flow statement]

Balance Sheet

[insert balance sheet]

❑ Key Assumptions And Risks

The company chapter of my business plan has the following sections:

❑ Company Overview

❑ The Team

The financial plan chapter of my business plan has the following sections:

- ❑ Sales Forecast
- ❑ Personnel Plan
- ❑ Profit And Loss Statement
- ❑ Cash Flow Statement
- ❑ Balance Sheet

CHAPTER SEVEN

Growing your Business and Leveling up

As the internet and digital business technologies continue to lower entry barriers in many sectors, business owners worldwide face increased competition and noise.

while, most of the market poses a little challenge, one or two firms still seem to emerge from the shadows and effectively steal clients. What will modern companies do to maintain their competitiveness?

Enter Specialization

Choose an industry and investigate the landscape. There's a good chance that there's a lot of disruption going on right now.

New technology, shifting perspectives, and shifting economic conditions all play a role.

"When you look at the Fortune 1000 over the last 40 years, beginning in 1973, you can see that significant improvements have occurred," says Rufus Franck, founder of Consultants 500.

"By 1983, one-third of these firms had fallen off the list. By 2013, only 30% of the initial companies remained on the list. This rate of transition will accelerate, as only one-third of today's large corporations are expected to survive the next 25 years."

It should, however, leave you looking for a solution. You must respond to increased competition and rapid evolution. That is why so many companies are focusing on specialization.

"The consequences of specialization are such a large phenomenon that it affects almost everything in our everyday lives," admits Franck. "Our innate need for specialization is also one of the primary reasons for the existence of skilled services. Since society and companies cannot know anything, we need the assistance and guidance of professionals with practical experience."

Although you may not be emotionally prepared to consider breaking away from your generalized emphasis and taking a leap of faith to seek specialization right now, it is something you will need to think about in the coming months and years.

This is a sobering glimpse into the future. Still, it should act as a wake-up call for all business owners and entrepreneurs. Suppose, Fortune 1000 companies – those with vast resources and cutting-edge technologies – struggle to stay afloat over the next quarter-century. Where does that leave your (likely smaller) business?

Amazing Advantages Of Specialization

Why should you hone your skills? Assume all is going swimmingly and business is booming. Doesn't it make sense to keep things as they are and reap the benefits of a flourishing business?

Maintaining the status quo will help you for a short time, but it will not last long. And the world's most popular companies, from Amazon and Walmart to Netflix and Apple, are forced to adapt to the ebbs and flows of the marketplace.

Currently, the market clearly demonstrates that it rewards companies that exclude the fluff and focus on delivering quality services.

Here are a few of the basic advantages of specialization:

1. Improved Value Proposition

When you specialize, you can have a better value proposition to your target market than companies that generalize in a similar area. As compared to the other way around, you basically become a larger fish in a smaller pond.

"You become an expert in a particular area by specializing," says digital strategist, Paul Boag.

"This reduces the number of other businesses that can compete with you for the specialized job. Furthermore, you can have resources that cannot be replicated by automated tools or that justify becoming a permanent member of an in-house team."

2. **Reduced Learning Curve**

A lot goes into running a successful company, including keeping current on emerging trends, best practices, and innovations. If you work in a field where progress occurs quickly, such as medicine or IT, you are all too familiar with this challenge.

You will lighten the load by reducing the learning curve by specializing. It would be easier to concentrate your time and energy if you limit yourself to one or two niches. As a result, understanding, and skills improve.

3. **Increased Authority Perception**

If you have a brain tumor, do you believe a general surgeon or a cranial neurologist specialized in diagnosing and treating spinal cord tumors is more authoritative and knowledgeable? That is clearly a rhetorical query. The person with the specialty is regarded as authoritative.

When the company wishes to specialize, it gains a greater sense of authority in the marketplace. This allows you to charge a more fee while accepting fewer clients.

4. Increased Conversions

In general, specialization leads to higher conversion rates. Assume a customer searches online for a company to remove a virus that has infected her computer. During her scan, she comes across four different company names:

Although all four companies might assist the customer in removing the virus from her device, the fourth is more appealing. As a result, this business draws more buyers and achieves higher conversion rates.

5. Improved Networking

Finally, specialization leads to more opportunities for networking. This is actually a product of the circles your company travels in. When you specialize, you will notice an increase in word-of-mouth marketing (assuming you produce a high-quality product or service). You'll probably notice that you're communicating with individuals and organizations, that are related to your field. This increases the chances of forming profitable partnerships.

In practice, specialization is often a normal byproduct of routine business operations. In other words, you'll unintentionally discover small niches in which your company could benefit from specializing. Rarely, you can simply say, "I want to specialize in this."

"If you work in the same industry sector or on the same issue, a same narrow problem all the time, you can gain a lot of experience and knowledge in that specific area," says business consultant Ian Brodie. "Of course, don't overlook outside insight and experience since that's where major breakthroughs always come from. However, by working in the same area repeatedly, you will easily become an expert in that field."

Here are a few examples of natural specialization evolution:

- Buyer's agent to real estate agent
- From the general auto mechanic shop to the fast oil change shop
- Marketing consultant to business consultant

In nearly all of these cases, the company (or a professional) was already doing what they eventually specialized in. They learned enough to specialize as a result of their extensive experience.

Nobody predicts that your company will go bankrupt in the next 25 years; but, if the statistic that two-thirds of large corporations will fail in the next 25 years is right, you should be searching for ways to modernize.

One of the most important things you can do is specialize your company and concentrate on bringing value to your customers. A specialist business in a booming market fails far less often than a general business in a stagnant one.

Once you have figured your specialization and have started your business, the next ideal step would be to start growing your business, right?

Healthy And Efficient Ways Of Growing A Business

It is critical for your company's success to generate new business by expanding your customer base. It can, however, be extremely difficult at times.

Here are some helpful hints for expanding your customer base.

1. **Get To Know Your Clients**

Understand your customers' needs and create goods and services to satisfy those needs. You will learn more about your customers by personalizing your offerings and empowering them to provide feedback.

2. **Provide Stellar Customer Support**

Make sure your customer service is outstanding, and go the extra mile wherever possible. Your clients will not only remember great service, but they will also be more likely to recommend others to you.

3. **Cater To Current Customers While Looking For New Ones**

Put in place strategies to nurture current customers, such as keeping in touch with them through an e-newsletter or informing them ahead of time about promotional events. Around the same time, look for ways to get more work and expand your customer base. Make sure you strike the right balance between cultivating existing customers and finding new ones. Suppose you own or run an Aboriginal company. In that case, you can support it by taking advantage of a free listing in the Aboriginal Business Directory Western Australia (ABDWA).

A growing number of state government and private organizations are looking to partner with Aboriginal companies and use the directory to find suitable suppliers.

4. Use Social Media

Social networking is a powerful tool for promoting your company to potential clients and gaining useful feedback through 'social listening.'

Via social listening, you can learn what your customers are thinking about you, gain insight into their behavior, recognize keywords and patterns that relate to your target market and thereby enhance your customer service. Social media will help you develop your company's profile and attract new customers.

5. Attend In Networking Activities

Invest time in developing your networks – it is not what you know but who you know.

Networking helps you form relationships with others and refer customers to you by word of mouth.

6. **Host events**

Hosting your own event can be a perfect way to get to know your customers and develop relationships.

Invite some of your best current customers and invite them to bring their mates.

7. **Do Some Voluntary Work**

Growing brand recognition in your group is an excellent way to draw new customers. Consider sponsoring or engaging in a group event to lift the company's profile.

8. **Keep Tracking Your Progress And Improve Your Strategies**

You should monitor where your customers are coming from to determine whether your marketing efforts are successful. Don't be afraid to try something new. If something isn't working, change the strategy and devote more time to the tasks that produce the best results.

9. **Investigate Your Competition**

When going to market, if you really want to get your bid in front of the masses, you must first study the competition. Frasier claims to perform his study on two websites. The first is called Similar Web. AdBeat is the other. Both serve as sources of competitive intelligence.

It's your chance to insert x-ray lenses into all landing pages, ad copy, and funnel points.

This enables you to learn about every advertiser's online strategy. Find the commercials that have been running the longest and try to imitate them. That is the most efficient way to scale any company. If it has been proven and is successful with your rivals, it will likely work for you.

10. Organize A Customer Loyalty Program

Loyalty services are excellent ways to boost revenue. It can cost up to three times as much to attract new customers than to sell to a current customer. Other sources place this figure somewhere between four and ten times higher. However, no matter how you slice it, attracting new customers is costly.

According to Frasier, creating a customer loyalty program would help you attract customers. It can also assist you in attracting new customers. It will pay off in the long run if there is a real reason to spend more money with you. Create an appealing loyalty program, make it available to your current customers, and watch your profits skyrocket over time.

11. Explore New Prospects

Analyze emerging market prospects by better knowing the population. Understand everything from distribution networks to direct rivals, as well as an international market and other future industries research. With the proper amount of research, there are certainly hundreds of new opportunities you might explore right away.

12. Create An Email List

Building an email list is one of the easiest and most important ways to quickly develop a company. Obviously, this necessitates the use of a lead magnet. Why else will people join the mailing list? A lead magnet necessitates the use of a sales funnel. Consider Aweber, ConstantContact, ConvertKit, Drip, GetResponse, and other list-building and management services.

13. Form Alliances

Strategic alliances with the right companies will really make all the difference. It can allow you to quickly reach a large number of customers. Identifying such collaborations can be better said than done.

However, keep an eye out for businesses that are complementary to yours. Contact them and suggest ways for you to collaborate.

Ways To Improve Your Business

Once you have developed a dedicated fan base, the next step is to explore how you can improve your business.

1. **Develop An Identity For Your Brand**

You do not have the money to have a "profile" for your small business, unlike enterprise-level companies. You have something better... YOU. I understand what you're saying. How do you deal with charismatic and well-known individuals? Simply be yourself. That is what the customers want. They want someone with whom they can relate and engage because you are genuine. What distinguishes you from other small business owners is your honesty, spirit, ideals, and vision.

That may be the most valuable resource you have. Simply put, develop the personality of your brand by incorporating more "YOU" into your company. The simplest way to do this is to simply be yourself while writing a blog post, filming an instructional video, or speaking at an industry function. Share your vision and principles on your "About Page," or write an editorial explaining why you started your company.

2. Established A Good Structure

This isn't as fun as creating your brand. However, these procedures have the potential to make or break your small company.

As described by Robert A. Norman on Business Know-How,

"A strategic business plan is your company's roadmap that explains your company's concept, philosophy, and purpose".

An organizational structure is the set of policies and procedures that ensure the company runs smoothly, such as job duties and discipline.

Operational Support Systems can relieve management of day-to-day tasks such as meeting scheduling and cash flow monitoring. This helps you to concentrate on tasks that will help your small business expand.

These three aspects will give you a better understanding of your company's processes and ensure you have the proper structures in place. According to Dr. W. Edwards Deming,

"85% of the reasons for failure to meet consumer needs are due to failures in structures and procedures — rather than the employee."

3. Avoid Common Mistakes

Learning from the mistakes of others is one of the easiest ways to improve your chances of success. So, here are four of the most common mistakes made by small business owners:

- *Your product or service's demand is overestimated.* Assume you continue to receive DVDs from Netflix in the mail. Just because you can't let go of this now-outdated technology doesn't mean you should start a DVD rental company. Your chances of survival will be slim to none between Redbox and streaming services. Test your ideas with friends and family, or perform primary market research.

- *Making a foray into a dynamic market.* On the one side, you know there is a demand. However, the demand could be oversaturated. This only works if you have a distinct competitive advantage.
- *Not accounting for unnoticed expenses.* Expenses such as your wages and the purchase of an office and supplies are self-evident. However, numerous costs are often ignored when starting a company, such as these 25. Not accounting for these costs which result in underutilization of resources or throwing your budget out of whack by purchasing items you do not need.
- *Profitability is not planned for.* If you want to be successful, you must understand your profit model. This entails being mindful of the gross profit on revenue, net margin, and the amount required to break even. These KPI-based numbers will show you how well your company is doing.

4. Get To Know Other Business

Customers like the fact that they can help local businesses. After all, by supporting a local small business, they are helping to improve the local economy. Knowing who is behind the product or service results in improved customer service and a more personalized experience. Furthermore, supporting small businesses helps to keep the neighborhood distinct.

As a small business owner, you can look for ways to help others by volunteering, hosting a fundraiser campaign, or supporting a non-profit organization. Participate in neighborhood activities and simply engage the community by soliciting input or showcasing your loyal customers. In other words, get active in the community and watch your company flourish.

5. Take Care Of Your Employees

"The workers are critical because their skills are what keep your computer running," says Gary Vaynerchuk. Vaynerchuk has been able to scale up his companies and "build dedicated teams while we continue to innovate" by prioritizing employee satisfaction and well-being.

Gary, like other great leaders, cultivates genuine relationships with his workers and promotes work-life balance. I just use remote teams for my companies. This gives them autonomy and versatility. I often inspire them by congratulating them on well-executed tasks, fostering confidence, and demonstrating to my team that the work they contribute serves a greater purpose.

Regardless of how you prioritize your staff, they will be your most ardent supporters, allowing your company to expand at a faster pace.

6. Try To Get Long-Term Customers

Your clients are your company's lifeblood. As a result, you must obtain and maintain them by:

- Advertise your products or services to your target audience through local newspapers, TV/radio advertisements, direct mail, trade magazines, directory ads, internet lists, event or charitable activities, or supplier co-op ads.
- Discount coupons, free trials, and promotional swag are examples of promotions and premiums.
- Current referral services provide an opportunity for referring your company to someone else.
- Providing excellent customer service.
- Consider Apple in terms of creating a brand that consumers choose to help.

7. Keep Yourself Organized

Being structured increases the chances of completing assignments and staying on top of all that needs to be accomplished. Making a regular to-do list is a good place to start. These are the most important jobs, which you check off when you complete them. It's a quick and easy way to keep focused and avoid missing something.

8. Make Your Business Standout

Do you want to stay ahead of the competition? Then, become an authority on a specific subject and target a specific niche within an existing industry. Does that sound complicated? You can begin by doing the following:

- Creating a definition for your business or knowledge base.

- dividing a large market into smaller niches
- Use the SPAN approach to put your expertise to the test (Subtopics, Pain, Attainable, Numbers).
- Using material, you can become an educator and advocate in your niche.
- Write guest posts to get the material in front of new people.

9. Always Monitor Your Cashflow

According to the results of "In Search of Solid Ground: Understanding the Financial Vulnerabilities of Microbusiness Owners," the true key to small business success is "the owners' discipline in handling their personal finances and business cash flow."

Take the following steps to ensure that your company has a healthy cash flow:

- Adjust your spending.
- Accelerate the processing of receivables.
- Increase the duration of your payables calendar.
- Make necessary adjustments to inventory levels.
- When you need immediate cash flow, consider obtaining a bridge loan.

Create a cash flow plan for the next six to eight weeks, so that you can deal with any cash flow problems before they become a problem.

10. Explore Management and Credit Availability

You would not want to take out a line of credit in an ideal world. However, there will be periods that you will need a small business line of credit, such as when you need to bridge a short-term gap between receivables and payables due to seasonality, inventory expansion, or business development. Perhaps, an emergency occurs, such as machinery failing, and you do not have the cash on hand to repair it.

Whatever the case may be, you must have access to credit when you need it. Furthermore, having good credit allows you to receive more favorable payment terms, have a payables float, and create new vendor relationships.

11. Build A Street Cred

Being a member of the group is essential to the success of your small company. After all, when you reach a new geographic market, you must first gain the confidence of existing businesses in the region. In addition to being involved in the group, use testimonials, word-of-mouth endorsements, certifications, and professional presentations to establish credibility.

12. Understand what you stand for.

TOMS creator Blake Mycoskie once said, "The easier it is for someone to understand who you are and what you stand for, the easier it is for that person to spread the word to others."

As an example, consider TOMS. The concept is extremely basic. Every time you make a purchase, the business will assist someone in need. You are well aware of who they are, what they stand for, and what they do.

Determine what you stand for from the start and tell your story in a way that is easy for your customers to remember and share. Again, it all comes back to developing your brand.

The final thing, according to me, that you can do to level up your business is trying out some inbound marketing.

In this modern-day and age where everything is going digital, Inbound marketing will do your business wonders to gather more potential customers and clients.

What Is Inbound Marketing?

Inbound marketing is a form of marketing focused on creating and delivering valuable content to your customers. The main goal is to create content that your customers can relate to or that can actually solve their problems to motivate them to visit your website, purchase a product or make a booking. This method focuses on growing your business through carefully tailored campaigns that will attract the attention of your target audience and cause a certain emotion or offer a solution and help build a relationship that will eventually convert to sales. It is the method of not forcing the actual product ad on your potential customers, rather waiting for them to come and find you.

Have you ever left your email address in exchange for a free file? Maybe you decided to subscribe to your favorite brand's newsletter to enjoy free shipping or discount on your next order?

Whenever, you gave your email or personal information to a business in exchange for some free value or a perk, you have actually entered an inbound marketing funnel. The idea of receiving something in return for their email address is becoming more acceptable by consumers. So much that if a person likes a brand, they will most likely see if there is any benefit from subscribing to their newsletter or becoming a member.

We should reflect on the traditional marketing concept, also known as **Outbound marketing**, to fully understand the concept of inbound marketing. For years, businesses have been pouring all they have into outbound strategies to "push" their ads in front of the general public.

As outbound marketing heavily relies on platforms such as TV, radio, print, and street billboards and, nowadays, some paid advertising and digital channels, they present a convenient way to get your message in front of a wide audience.

This concept can also utilize the power of trade shows, cold calling, and seminars to get the message out to as many potential customers as possible.

The Basics Steps Of Inbound Marketing

Inbound marketing is all about attracting new customers to your business by creating industry-relevant content that is both engaging and informative. We have already touched on how this content can be shared in blogs, social media posts, Instagram stories, or live videos, to name a few. The options are endless, as inbound marketing allows you to adjust your strategy and content type to your main business operations.

By combining multiple types of content across social networks and other digital channels, you will raise your brand awareness while offering solutions and relatable content to your target customer.

1. Attract
2. Convert
3. Close
4. Delight

The first step is to attract the right customer and then focus on converting them to a lead, closing the sale, and keeping in touch with them by providing remarkable and fresh content that will lead them to recommend your business to others.

To summarize, inbound marketing has proven to be more effective when compared to traditional outbound strategies as it is focused on reaching your potential buyers organically. So, how do you start your inbound marketing funnel?

Step 1 - Attract the right audience

The first step is to attract the right audience. This means finding those prospective customers interested in your brand and are likely to make a purchase. The key is to create content around the needs and potential problems that your ideal buyer might have, which can be done through multiple channels.

Blogging is one of the most effective ways to attract your desired audience. Why? The answer is simple - blog posts allow you to share informative and educational content that customers find helpful and are most likely to engage with. Simply having a blog on your website is not enough. For your customers to find you, you need to increase your brand awareness or increase your chances of being seen online.

Think about a brand that you like. How did you find about them? The main concept of inbound marketing is built on three main stages:

- Did you read something on their blog?
- Watched a video from their YouTube channel?
- Or maybe came across a live session that they hosted on social media?

Every business has some form of value to offer customers, no matter what industry. Your job is to identify an end goal and turn it into an engaging piece of content that consumers will want to interact with. A great example would be blog posts or videos reflecting on hints, tricks, and tips related to your industry or a certain niche. You may share FAQs or guides, and templates that your customers can benefit from.

Step 2 - Convert Visitors into Leads

Now that you have your visitors' attention, your goal is to convert them into actual leads. This is mostly done through newsletter signups. They allow you to collect their personal details and add them to your database. You will have to offer some content for exchange, usually in the form of an e-book, *"cheat sheets,"* or even infographics. The idea is to link a free downloadable file to help your customer resolve a problem or benefit from it.

For example, if you are in the e-commerce business, you might offer 10% off their first purchase, if they subscribe to your newsletter. A **Service-based business** such as a design agency might offer a free mood board template to those who join their email list. Either way, the goal is to build a long-term relationship and convert your visitor by gaining their email address, allowing you to send your promotions, news, and product releases straight to their inbox. And, since they voluntarily left their details, they confirmed that they are interested in your business, so you know that you have a valid lead who is actually interested in purchasing from you.

That is the power of inbound marketing. Offer your products and services to those who want to hear about your business instead of intrusive advertisements shown to the general public, hoping that those interested will respond.

Step 3 - Close the Deal

It is time to add one happier customer to your list. Now that your visitor is officially a lead or prospective buyer, it is time to show them your offer and score an actual sale. Closing the deal is all about looking into your leads, analytics, and sales performance to identify what drives your leads to spend their money on your website.

The first step is to identify how *"qualified"* your leads are or - how ready they are to make a purchase. The best practice to identify qualified leads is lead scoring. You need to compare your leads and their actions to identify how they are to purchase your product.

A good way to analyze the behavior of your leads on your website is to check your analytics and see what your visitors search while they are on your website. You could track search results and look into the products added to the cart the most. Knowing these will help you merchandise your website and ensure that best-selling and most wanted products are displayed in a highly visible spot as they are the ones driving sales and actually making you money. So, next time a qualified lead visits your website, they will instantly be able to see your best offers.

Your leads may require a bit of motivation to pursue their shopping journey and successfully check out. When it comes to online stores, free shipping over a certain threshold is still among the main conversion rate drivers. It is considered rewarding as customers usually prefer to spend an additional £5 or £15 instead of paying £10 for shipping.

Another factor to consider is the abandoned cart rate or the reason why leads decide to leave your website without finalizing their transaction. Sometimes, buyers just need a little reminder that amazing products are waiting for them in the cart. A popular way to handle abandoned checkouts is to send email reminders **1-2** hours after leaving.

Not all consumers are driven by the same factors, and knowing what drives sales on your website can help you convert leads faster. Inbound marketing allows you to tailor your strategy to suit one or many buyer personas, which is especially important when closing a sale. This is why tracking and analyzing your website performance regularly is always helpful.

Step 4 - Delight Your New Customer

Inbound marketing is all about building relationships, so it heavily relies on "surprise and delight" strategies. These types of strategies are focused on keeping the customer happy even after the purchase was made. By ensuring that your small business provides a maximum customer support level throughout the process, your customers will start to build trust in your business and prefer your services over competitors.

A great way to provide support and build a strong after-sale system is to utilize website live chat where your customer service team can provide prompt solutions to your customers.

Imagine that a customer ordered one of your products, paid for the order but never received a shipping confirmation email. System failures and glitches often happen. And most of the time, they have nothing to do with your business. However, since you provide the service to your customer, your business is the one they will be upset with.

After placing this order, days go by, and the customer realizes that the package has not arrived yet. By having the option to simply message your team on chat, this

customer would avoid long reply times and get her query solved within minutes. This could save you from negative reviews, refund requests as well as complaints, and negative feedback on social media. Your instant availability to solve the missing parcel problem will increase your credibility in your customer's eyes. The customer will feel safe and encouraged to shop at your online store again.

Another way to delight your new customer is to use email automation and ensure that relevant information is delivered straight to their inbox. Once the product or service is delivered, you could use email as a tool to ask your customer for feedback, keep in touch with new products or services. After all, they already know that your product or service works. Offer them premium content and show them that they are valued as a customer.

You can also send a quick survey for them to fill. This insight can greatly benefit future campaigns, help understand how your customer perceives your business, and why they trust you. It is also a great way to find out about any negative experience. You can use this feedback to develop content and educate your customers on potential troubleshooting problems and ways to get in touch and resolve it.

The Channels Of Inbound Marketing

By allowing you to reach a highly targeted audience, inbound marketing relies on different channels and content formats. As mentioned, there are plenty of inbound marketing channels. The choice on which one to pursue should always be based on the nature of your business. The rule of thumb is to go where your audience is. That will help your content be seen by the exact consumer group you aim to convert. So, to identify the right type of inbound marketing channel for your business, let's reflect on the main channels used by small businesses in their inbound marketing strategies.

1. Search Engine Optimization

It all starts with Search Engine Optimization, commonly referred to as SEO. Suppose your content is not optimized for search engines. In that case, chances are you will not attract any organic visitors to your website. By implementing effective SEO techniques, you allow the search engines to communicate with your website and show customers that you have what they need. However, search engines cannot completely understand or comprehend your website content. This is why they rely on SEO to find out how much your page is relevant to a customer's search query.

SEO is one of the most important inbound marketing channels. It boosts organic reach and decreases costs and the need for paid advertising. To properly SEO-proof your online business, it is important to understand that there is no 'one size fits all approach. Every SEO strategy is tailored to the needs of individual businesses. It needs constant attention and tweaking to provide desired results.

The main focus of SEO is on free, organic traffic, and it takes a lot of hard work and research. Top organic ranking spots in Google search results cannot be paid or bided on. They are earned by businesses and websites that have nailed their SEO strategy, both on-page and off-page.

On-Page SEO is focused on optimizing individual pages of a website in order the help them rank better. It pays close attention to both written and visual content and HTML source code to help search engines determine on-page relevancy.

Things to consider when improving your on-page SEO are quality content, primary, long-tail keywords, title tag and meta descriptions, and alt text and URL structure. Off-Page SEO, however, refers to those ranking factors that happen off your website. It is mostly focused on backlinks to other sites, domain authority, local SEO such as Google My Business. Working on your complete SEO strategy will help you set the ground for a successful inbound marketing campaign, regardless of the industry you are in.

2. **Blog**

As one of the most incorporated inbound marketing tools, a blog is an essential part of your marketing campaign. It allows you to provide valuable content, promote your products or services and inform your customers of any industry news, changes or improvements. Having a business blog helps you build your brand image and educate your target audience on what your business is all about.

When it comes to inbound marketing campaigns, a blog is the essential part of content creation as it is the most common platform where user-friendly content is shared.

From articles to videos, photos, and infographics, a business blog allows you to connect with the audience that resembles your buyer persona, attract their attention to your band and eventually, turn them into a paying customer.

As one of the leading marketing channels, blogging is beneficial to your SEO strategy and brand reliability. Your blog has a positive impact on the search engine's index by publishing regular and fresh content. It brings up to 55% more visitors and 97% inbound links, according to Hotspot statistics.6 The most important benefit of having a blog is that it drives more traffic towards your business. That means more leads that you could convert into actual customers. Therefore, a blog is a perfect spot for sharing your inbound marketing content and attracting new leads, and gaining greater exposure. A blog also allows you to share your content on Social Media and drive traffic from new visitors and potential leads who are eager to learn more about your brand.

To get the most out of your blog in an inbound marketing campaign, it is important to create a posting schedule and stick to it. Knowing when and what to post may make or break your marketing campaigns. Creating a posting schedule for your business comes down to planning when and where you need to post your content to stay on top of your strategy. The best practice is to review your analytics and see when your audience is online and when are your peak engagement hours and then use this data to plan out your posting schedule.

For example, suppose your data shows that your audience is online during peak hours of the day when most people travel to and from work while browsing their social media. In that case, you can plan to post around that time to increase your chances of your content being seen. This will help build up your domain authority and increase your exposure.

3. Social Media

Social media is among the top inbound marketing channels. It allows you to instantly connect with your followers and share quality content. Any small business can highly benefit from social networks, especially if you are a product-based business. Social networks are evolved and developed into one of the most convertible marketing channels. Their structure is shifting from networking to sales-oriented platforms. One by one, social media networks are turning their feeds into mini-stores and developing an engaging environment where customers are encouraged to buy the products they like and share their feedback, and interact directly with brands.

This has contributed to the birth of the influencer trend. Leaders in their niche have a strong following and influence are collaborating with brands to promote their products to a wider audience. This is especially useful in the fashion and beauty industry and can play a major role in the overall inbound marketing strategy.

Imagine having a makeup line business where your products are sold on your website. You could blog or make video tutorials to promote your products to your buyer persona, or you could find a beauty influencer who shares the same values as your ideal customer and ask them to collaborate. Let's say they make a quick tutorial using your products and publish it on their account while also giving you the approval to use it on your social accounts. By tagging you and mentioning that they partner with your brand, you would gain attention to their large following. In this case, the influencer would use their talent and exposure to attract leads to your website, where you can continue your funnel and offer them a discount in exchange for their email. From there, you are back, focusing on closing the deal and delighting the customer.

Many large brands have used this trend on social networks and built strong relationships with consumers and influencers. These collaborations with influencers resulted in increased traffic and sales.

4. Email

Email marketing has remained one of the most popular channels to build one-on-one relationships with customers and subscribers. It is the perfect channel to share personalized content and show your subscribers that you care. It is the casual conversation between you and your existing or potential customer. You can share some form of value. For instance, just before the holiday season, you can use your newsletter to share gift ideas for the whole family.

Your emails should be crafted in a friendly, casual yet informative way and focused on the reader's needs. Best practices suggest keeping the message brief and occasionally using this form of marketing as you do not want to spam their inbox. Remember, they chose to leave their email as they trusted your business, so do not disappoint your subscriber with content sale offers. However, do include a CTA or a Call-To-Action in the first or second paragraph.

A CTA is a popular marketing term used to prompt a response or reaction that will result in an immediate sale. The most used forms are 'Shop Now, 'Order Now,' but they can always be adjusted to suit the main purpose of your email, such as `Shop Christmas Gifts' or 'Shop New Arrivals.' The point is to give your readers a clear indication of what actions you want them to take

Many large brands have used this trend on social networks and built strong relationships with consumers and influencers. These collaborations with influencers resulted in increased traffic and sales.

Emails are used by almost every industry, and their content should be carefully planned around the buyer persona's needs. You could also use this form of communication in the delight phase of your inbound marketing strategy and offer birthday discounts, upsells, or a special promotion for existing subscribers.

5. Video

It is believed that video content can boost your visibility both in search engine results and on social media: These days, everything is about the video. It is one of the fastest-growing content formats, mostly due to advanced technology on mobile phones and faster Wi-Fi connections, which allow more and more people to stream videos without worrying if they will lose all their data in a day.

Videos are used across all inbound marketing channels and platforms; they can easily be embedded in your landing page or blog posts, posted to your social media feed or story as well as featured on YouTube. They are also a great way to boost your SEO as videos published on YouTube are also featured in Google search results. Video content may sound demanding or hard to film. Still, in reality, all you need is a script, natural light, and a solid phone camera. Any additional equipment is always beneficial, but for the basic video formats, the basics will do.

In **2020,** Tok-tok has joined the trend of popular social networks and especially became popular among the small business sectors. Soon after, Instagram introduced

'Reels," allowing you to create and publish another video format to your feed. These short videos are a great way to connect with your ideal customer and potential leads. You can use them to showcase what is happening behind the scenes on a product photoshoot, show off your works in progress, share industry tips and tricks or demonstrate the popular ways to wear or use your products. Options are endless, and you have maximum creative freedom.

6. **Podcasts**

Although podcasts were around for years, their popularity has significantly increased in recent years, especially in the business-related sector. Podcasts could be a great way to offer valuable content through interviews with other recognized members of the industry or discuss specific problems that your buyer persona or ideal customer faces. Podcasts are especially beneficial to B2B space. Still, depending on your industry, this channel could be further analyzed to see if there is any potential to create engaging content for tour audiences.

CHAPTER EIGHT

Understanding the Importance of Choosing a Business Location

The final chapter will be a brief, albeit extremely important, one. It will discuss the importance of the right location for your business and the factors that you should keep in mind.

One of the most critical aspects of starting a small business is deciding on a spot. Finding the right location is critical, particularly if you run a small retail or restaurant business. It can influence foot traffic, the business climate, and your small business's long-term success.

Finding the right location entails knowing what characteristics to look for in a potential room. Analyzing your place, reading about potential consumer demographics, and considering where competitors are located are critical aspects of locating and selecting the right spot.

Before deciding on a venue, you should also understand your needs as a business owner. Suppose you're not running a retail or restaurant-type business and therefore won't need to rely on foot traffic. In that case, you should think about access for other business needs. This may include stock deliveries, client meetings, and warehouse or large storage space for your business. You're well on your way to finding the right business room for you to analyze your choices and define your needs.

Why Is A Location Important?

Suppose you own an outward-facing business in some way, such as dealing with clients or customers. In that case, you must carefully consider how your location can benefit your company.

Depending on the type of company you run, your location should be welcoming and refined. It should form a favorable impression in the minds of your clients and customers, causing them to regard your company favorably. [Related Article: What You Should Know About Choosing Retail Space]

When contemplating a new business site, there are some crucial steps to take. There is more to locating the ideal location than simply locating an available structure for your company to inhabit. Your organization's position can be defined by how you choose it. Before exploring other variables, consider the following steps:

- Do your homework. Understanding local zoning codes, mapping rivals' positions, and keeping an eye on what properties are available are all part of the process of finding the best business spot (or may become available soon). You can also study your target market and any additional advantages or limitations provided by local or state agencies.
- Think about taxes. Think about the state and local tax consequences of your potential venue. Depending on the nature of your company and its venue, it could be more profitable to search out a location outside of a certain zone or municipality to obtain a tax break.
- Inquire about any government benefits. Small companies in some sectors may benefit from federal and state incentives. You can be eligible for certain offers if you choose the right location for your company.

Your location is critical to your success. It is your company's external appearance. It sends a clear message to your customers about the type of business you operate.

What To Look For When Choosing The Business Spot

You can find the right location for your company by considering your business, what's available in your general area, and what types of features are relevant to you. Of course, affordability is often the most significant aspect. If you're looking for the perfect location for your company, it's important to weigh your options and work with a local bank.

Aside from affordability and loan conditions, there are other considerations and questions to consider when looking for the right business spot.

1. What is the significance of location in your business?

This may seem like a no-brainer, but remember how the location will affect your company. The location of a retail store differs greatly from that of the headquarters of a central office. When determining the effect location can have on your company, consider your customers and clients, as well as your employees. You want to choose a venue that is convenient for everybody.

2. Will You Be Able To Afford It?

Until you begin searching for a business site, you must first build a business budget for your upcoming expenses. Since you are obviously not flush with cash as a new company, your budget is one of the most critical business location considerations. How much money would you spend on a business location?

Keep in mind that the monthly rent or lease payment does not reflect the overall cost of the location. There are additional costs associated with the house.

Find out if you have to pay taxes in your area if you buy a commercial house. Consider the burden of property taxes as well. The amount you pay in taxes can vary depending on where you live and whether you buy outright or rent.

The overall cost of services should be factored into your business location plan. Older buildings can be more expensive to heat and cool. You would be able to find out how much the services cost per month from the previous owner.

There may be additional costs associated with the construction. You can need to renovate or add improvements. You might need to install insulation or run communications cables, for example. You'll also need to decorate the interior to represent your personal style and brand. The scale of the required building changes will be determined by the age and previous use of the building.

3. **Is The Space Sufficient?**

What kind of room do you require? Determine how you plan to use the business spot. Can you ever use it for shopping purposes? Do you need a workshop or an office? Do you want a storage facility or a warehouse? Do you need a kitchen or a space for specialized equipment?

Knowing what sort of building you need can assist you in determining whether or not prospective buildings are a good match.

4. **Does It Fit Your Style?**

Certain small business locations can be ruled out based on the company's logo and visual style. Your brand and your structure should be consistent. For example, suppose your company sells retro superhero memorabilia. In that case, a colonial-style building with chandeliers may not be a good fit. You may be able to decorate the building to match your brand, but your contract could forbid you from doing so.

5. **Can the company continue to accept shipments of goods?**

Choose a place with warehouse storage space and convenient distribution choices for clients and consumers if you work in the industrial sector or operate a company that receives large supplies of products. A company that specializes in shipping and storing merchandise needs some structural facilities, such as loading docks.

6. **Will you have meetings at your place of business?**

Consider, how your clients and consumers will perceive your location once more. If you deal with customers and clients regularly, you'll need a convenient venue. It should also have reception areas, waiting rooms, and meeting rooms so that you can meet with a variety of clients and customers during the day.

7. **Will you, your clients, or your employees need specialized parking?**

When deciding on a business venue, keep your company's size in mind. Your workers will almost certainly need to drive to your place. Is there parking available if that's the case?

Similarly, if you deal with clients and consumers regularly, you'll need easy parking options for them. The parking options available around a business location are sometimes the most critical part of it.

8. **Who will be able to view and communicate with your location?**

Consider who will visit the company daily. You must design a place that satisfies both their needs and yours as a business owner.

When selecting a place for your company, you must consider many factors, including whether there is enough space for staff and customers, whether it is easily accessible for shipments, and whether customers can visit your business regularly.

Choosing The Best Location

Aside from the factors mentioned above, when it comes to deciding where your company will be located, several critical measures are taken to make the right decision.

You want to weigh all of your options and the viewpoints of as many people as possible who are involved with your company. Choosing the best business location necessitates doing as much analysis as possible.

Consider the following considerations when you look for the best location for your company:

Target Audience: Consider the target audience's demographics and where they live. Investigate various demographic aspects of your local area, especially in the vicinity of your desired location. If your company depends on foot traffic, you can try to put yourself as close to your target audience as possible. Knowing this information allows you to learn more about disposable income in your region and how high demand for your product or service might be.

Competitors: Consider the rivals' locations. You don't want to live next door or across the street from a competitor's company. Knowing where your rivals are will help you find a business place to build your own base. It can also help you better gauge demand for your product or service in comparison to other businesses.

Location: The building's structure is choosing the best business place is about more than just the land itself. Choose a place with a building that you and your employees will enjoy. Consider it from the perspective of a future client or customer. You'll need a place with a structure that works best for your company and its clients or customers.

Zoning: Look up zoning maps for your town or area. This helps you to determine what business practices are permitted in your local area. If you own a retail store or a restaurant, proximity to commercial zoning is critical. Because of nearby residential zoning, you can have direct access to foot traffic in your neighborhood.

Nearby businesses: Consider what other companies are nearby your potential site, just like you might consider where your competition is located. This is significant because those businesses will attract customers that fit your target demographic. When considering your neighbors, make an informed, balanced decision. You want to work with other companies to help develop a stable economic climate. You should not be forced to accept adversarial business relationships.

Foot traffic: For certain small businesses, high foot traffic may be the icing on the cake. When looking for the right spot, if it's in a high-trafficked area, go there many times during the day to see how crowds ebb and flow. Do your best to determine if your company will profit from the amount of foot traffic in the potential location's designated area.

Consider the demographics, what competition and other businesses are nearby, any zoning problems that might be a problem, and how much foot traffic you might expect before deciding on a spot.

Suppose you keep all of these in mind and practice making a checklist out of these before choosing a location. In that case, the chances of you choosing the most perfect spot will dramatically increase!

Please Leave a Review!

I would be incredibly thankful, if you could take just 60 seconds to write a brief review on Amazon. Even if it's just a few sentence!

CONCLUSION

As much as we would like to believe, smartly saving money and properly investing them in a business that will bear fruit, in the long run, isn't an easy thing to do. No matter how much you try and how much you stay strict on your path, life will always throw something unexpected to you and derail you from your path.

An unexpected family emergency, someone you love falling terminally ill, a sudden business trip, or even an unforeseen house repair! The list of things that can go wrong in life is almost limitless.

As a worst-case scenario, you can take the recent COVID-19 pandemic as an example. No one knew that this would come; it came out of nowhere and completely brought the global economy to its knees, causing serious financial issues for thousands of companies, agencies, businesses, and millions of individuals out there.

You won't ever be able to completely prepare yourself for the uncertainty. Still, with the methods and strategies outlined in this book, you will be able to take the right step towards creating a financial foundation where you are confident enough to tackle these issues head-on!

Let the ideas shared in this book act as a primer. You can start building up your financial empire through efficient money management and business developing strategies.

Throughout the pages of this book, I have shared the core importance of saving money, some awesome strategies to both save money and earn money using the most advanced and popular methods to date.

The aforementioned steps are designed to pull you out from your current debt and difficulties and make you confident enough to move on to the next phase, developing a business.

And keeping that in mind, the final chapters of the book completely focus on holding your hand and walking you through the core steps of how to properly research your market, how to draw the perfect business plan, and how to implement that idea and take action to turn the vision of your business into reality. I have also touched upon a concept that most books often ignore. That is the importance of choosing your business location, which is of paramount importance.

To top it off, Effective strategies and tips on how to grow your business further once established are also provided at the end of the book to help you establish you're into deeper territories and increase your reach.

I sincerely hope that you found the contents of this book to be valuable and helpful.

Whether you are struggling financially or looking to ensure a safe future, please, never lose hope! Engrave the concepts provided in this book in your heart and start implementing them in your life; I promise you that down the line, they will definitely serve you well and give you the financial freedom that you desire.

As a final request, I would encourage you to leave feedback for my book on Amazon if you found this book helpful and effective. It would mean a lot to me and inspire me to produce more high-quality content for you in the future!

Stay Safe and Stay Blessed.

Just For You!

A FREE GIFT TO OUR READERS

Ready-to-use templates that you can download and start mind mapping with right away! **Visit this Link**

https://infobusiness001us.activehosted.com/f/1

Made in the USA
Monee, IL
08 November 2026

67378333R10166